The Wild Orchids of Arizona and New Mexico

The Wild Orchids of Arizona and New Mexico

Ronald A. Coleman

COMSTOCK PUBLISHING ASSOCIATES
a division of CORNELL UNIVERSITY PRESS
ITHACA AND LONDON

Copyright © 2002 by Cornell University

All rights reserved. Except for brief quotations in a review, this book, or parts thereof, must not be reproduced in any form without permission in writing from the publisher. For information, address Cornell University Press, Sage House, 512 East State Street, Ithaca, New York 14850.

First published 2002 by Cornell University Press

Printed in China

Library of Congress Cataloging-in-Publication Data

The wild orchids of Arizona and New Mexico / Ronald A. Coleman.

Coleman, Ronald A.
 The wild orchids of Arizona and New Mexico / Ronald A. Coleman.
   p. cm.
 Includes bibliographical references (p.   ) and index.
   ISBN 0-8014-3950-7 (cloth : alk. paper)
   1. Orchids — Arizona — Identification. 2. Orchids — New Mexico — Identification. 3. Orchids — Arizona — Pictorial works. 4. Orchids — New Mexico — Pictorial works. I. Title.
    QK495.O64 C63 2002
    584'.4'09789 — dc21
                                                                    2001005090

Cornell University Press strives to use environmentally responsible suppliers and materials to the fullest extent possible in the publishing of its books. Such materials include vegetable-based, low-VOC inks and acid-free papers that are recycled, totally chlorine-free, or partly composed of nonwood fibers. For further information, visit our website at www.cornellpress.cornell.edu.

Cloth printing   10 9 8 7 6 5 4 3 2 1

To the memories of my father Cecil Coleman, my mother Bernice Coleman, and my sister Gaynell Coleman Buis.

Contents

List of Tables and Keys   ix
Preface   xi

Introduction   1
Calypso   19
Coeloglossum   27
Corallorhiza   35
Cypripedium   59
Epipactis   69
Goodyera   79
Hexalectris   91
Listera   113
Malaxis   123
Piperia   141
Platanthera   147
Schiedeella   185
Spiranthes   193
Stenorrhynchos   209

Appendix 1: Excluded Species   219
Appendix 2: Watch List of Species Not Yet Reported from These States   221
Appendix 3: Herbarium Collections Studied   223
Appendix 4: Distribution of Orchids in Arizona   225
Appendix 5: Distribution of Orchids in New Mexico   227
Appendix 6: The Counties of Arizona and New Mexico   233
Bibliography   235
Index of Scientific and Common Names   243

Tables and Keys

TABLES

1. Orchids at Range Limit in Either Arizona or New Mexico   3
2. Species Distribution by County in Arizona   6
3. Species Distribution by County in New Mexico   9
4. Number of Counties in Which Each Species Occurs   11
5. Blooming Seasons   12
6. Nature Conservancy Ranking System   15
7. Species Ranking for Arizona   15
8. New Mexico's Rare Orchids   16
9. *Platanthera* Characteristics   182

KEYS

Key to Orchid Genera   17
Key to the Species of *Corallorhiza*   36
Key to the Species of *Epipactis*   69
Key to the Species of *Goodyera*   79
Key to the Species of *Hexalectris*   92
Key to the Species of *Listera*   113
Key to the Species of *Malaxis*   123
Key to the Species of *Platanthera*   148
Key to the Species of *Spiranthes*   194

Tables and ...

Preface

Searching for wild orchids is becoming an increasingly popular pastime across the United States. Exploring fields, forests, bogs, and swamps, with the hope of seeing and photographing an orchid, is euphemistically called "orchid hunting." Although the only trophies the "hunters" bring home from this endeavor are memories and photographs, the hunt itself is challenging and requires knowledge of the prey. Wild orchids are hidden jewels in quite diverse habitats, and it is always a pleasure to discover one serendipitously or during a deliberate search.

Possibly because of the growing popularity of orchid hunting, and certainly contributing to it, several regional orchid floras have been published over the last two decades. The pace of publication increased in the last half of the 1990s. However, the single most significant work on North American orchids remains the monumental *Native Orchids of the United States and Canada* by Carlyle Luer, published in 1975. Earlier, Ames (1924) and Correll (1950) had written books on orchids in North America, but Luer updated the nomenclature, including species missed by the others, and described several taxa for the first time. His was also the first treatment of native North American orchids that included color photographs. The authors that followed have recognized the debt students of North American orchids owe to Luer. The more recent regional orchid floras expand on the work by Luer, usually adding current nomenclature, more precise range definitions, and additional species based on evolving scholarship. A list of these publications follows:

Wild Orchids of the Middle Atlantic States, by Oscar W. Gupton and Fred C. Swope. 1986.
Orchids of Ontario, by R. E. Whiting and P. M. Catling. 1986.
Missouri Orchids, by Bill Summers. 1987 [first published in 1981].
Orchids of the Western Great Lakes Region, by Frederick W. Case Jr. 1987 [revised from 1964 publication].
Wild Orchids of Arkansas, by Carl M. Slaughter. 1993.
Orchids of Indiana, by Michael A. Homoya. 1993.
Orchids of Minnesota, by Welby R. Smith. 1993.
The Wild Orchids of California, by Ronald A. Coleman. 1995.
The Orchids of the Ottawa District, by Joyce M. Reddoch and Allan H. Reddoch. 1997 [special issue of the *Canadian Field-Naturalist*].
The Orchids of Bruce and Grey, by the Bruce-Grey Plant Committee. 1997.
Wild Orchids of the Northeastern United States, by P. M. Brown. 1997.
Orchids of the Northeast, by William K. Chapman. 1997.
Wild Orchids of Texas, by Joe Liggo and Ann Orto Liggo. 1999.
Native Orchids of the Southern Appalachian Mountains, by Stanley L. Bentley. 2000.

I moved to Arizona in 1994, having spent most of the preceding 23 years studying native orchids in California. At the time of the move, I had completed *The Wild Orchids of California*, which was in the process of being published. Much to my delight, the orchid flora in Arizona and New Mexico was about as large and diverse as that of California. *The Wild Orchids of Arizona and New Mexico* is a natural extension of my love for native orchids and helps fill the literature void regarding orchids of the southwestern United States.

One of my goals during the preparation for this book was to find and photograph all the native orchid species within the boundaries of these two states. The search for the plants thus became not only an adventure but also an exercise in scientific detective work. I studied herbarium records and literature for historical locations and searched suitable habitats throughout the appropriate regions. I did not fully achieve this goal because I was unable to find either *Piperia unalascensis* or *Hexalectris nitida* in New Mexico. The color plates for all but these two species are from photographs taken in New Mexico or Arizona. The photographs of *P. unalascensis* were taken in Olympic National Park in Washington, and those of *H. nitida* were taken in Guadalupe National Park in Texas.

This work would not have been possible without the generous support of Lucinda McDade, director of the herbarium at the University of Arizona, and of Phil Jenkins, collection manager. McDade sponsored me as a visiting scholar at the university, a position that allowed me access to the research resources of that tremendous institution, including access to the herbarium on weekends. Jenkins arranged for loans of specimens from herbaria throughout the Southwest, greatly facilitating the laboratory part of my study. Jenkins also reviewed the manuscript, as did David Bertleson, Mark Dimmitt, William Jennings, and Larry Toolin.

Pronunciation guides are provided for the genus names and specific epithets. These are borrowed from material in the works by Hawkes (1965), Keenan (1998), and Homoya (1993).

The field work to support this study consumed just about every weekend between April and October over the last 6 years. I logged many miles both by car and on foot, relocating historical locations and searching for new ones. Frequent companions on these orchid jaunts were my wife Jan and our sons Joel and Troy. On many trips friends such as Joseph Welch, Antoinette Sagade, Larry Toolin, Mark Demmitt, and several members of the Arizona Native Plant Society accompanied me. Their contributions to this work are gratefully acknowledged. Many thanks to all those who helped, and to all who share an interest in native orchids: Happy orchid hunting!

RONALD A. COLEMAN

Tucson, Arizona

The Wild Orchids of Arizona and New Mexico

The more common form of C. *bulbosa* with white lip

C. *bulbosa* with lip the same color as sepals and petals

The leaf of C. *bulbosa* appears in late fall and fades shortly after blooming in spring

Calypso bulbosa blooming in pine/fir forest early in spring

Plate 1

The short-lived plants of *Coeloglossum viride* var. *virescens* in light to moderate shade

The long bracts on *C. viride* var. *virescens* partially obscure the flowers

The three-lobed lip of *C. viride* var. *virescens* helps identify the flowers

The ellipsoidal seed capsules of *C. viride* var. *virescens*

Plate 2

Corallorhiza maculata blooms early in spring on colored leafless stems

A typical wide-lip form of C. *maculata*

The narrow-lip form of C. *maculata* blooms later than the wide-lip form

A plant with the spot pattern typical of C. *maculata* var. *mexicana*

Yellow form of C. *maculata* with lightly spotted lip

Yellow form of *Corallorhiza maculata* with pure white lip

Nearly every flower of *C. maculata* sets fruit

A typical dark-colored form of *Corallorhiza striata*

A lightly colored form of *C. striata*

A nearly pure yellow form of *C. striata*

Stripes are apparent even on the capsules of *Corallorhiza striata*

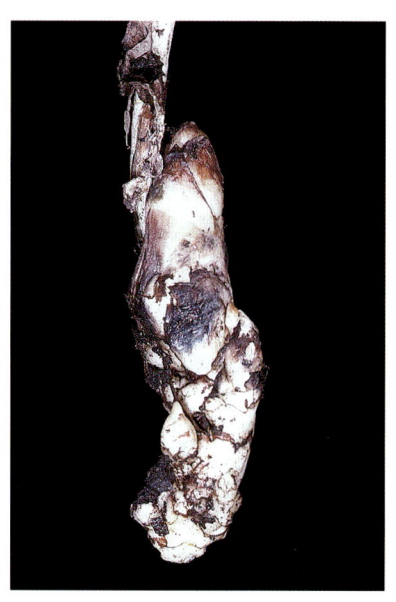

C. striata grows from a coral-like rhizome

Corallorhiza trifida grows in damp to dry woods

C. trifida with the pure white lip typical in the Southwest

Corallorhiza wisteriana in an open spot in conifer forest

The leafless stem of *C. wisteriana* blooms early in spring

C. wisteriana with finer, paler dots

Typical color form of *C. wisteriana* with rich purplish red dots

Corallorhiza wisteriana with pure white lip

Yellow-green form of *C. wisteriana* with pure white lip

Many capsules of *C. wisteriana* have a hint of green

Rare peloric form of *C. wisteriana*

The coralloid roots of *C. wisteriana*

Plate 7

Cypripedium parviflorum grows in moist to dry conditions

C. parviflorum has the largest orchid flower in the Southwest

Some plants of *C. parviflorum* have two distinct sepals instead of a synsepal

Capsules of *C. parviflorum*

The red-brown pigmentation is missing in some plants of *C. parviflorum*, so their sepals and petals are pale yellow or pale green

Plate 8

Epipactis gigantea in a typical streamside habitat

E. gigantea in a hanging garden

Large colorful flower of *E. gigantea* in Navajo County, Arizona

The hinged lip of *E. gigantea* leads to the common name of chatterbox

Epipactis helleborine is not native but is spreading in the Southwest

The central lobe of the lip on *E. helleborine* has a nectar reward for the pollinator

Most plants of G. *oblongifolia* have leaves with only a single white line, but some are highly reticulated

Goodyera oblongifolia grows in slightly damp to dry spots in the forest

Almost all flowers of G. *oblongifolia* set fruit

The sepals on G. *oblongifolia* are tan to greenish

The densely clustered tubular flowers of G. *oblongifolia* are easy to identify

Most plants of *G. repens* in the Southwest have unmarked leaves

Goodyera repens blooms in slightly damp places in the forest

A few widely scattered plants of *G. repens* are highly marked

Though tightly packed, the flowering stems of *G. repens* are much shorter than those of *G. oblongifolia*

The upright, ellipsoidal capsules of *G. repens*

The pure white flowers of *G. repens* help identify it

The shiny, waxy flower of *H. nitida* (photographed in Texas)

Hexalectris nitida is extremely rare in New Mexico (photographed in Texas)

Hexalectris revoluta grows in isolated, oak-lined canyons

Spike of *H. revoluta* under oaks in Pima County, Arizona

The tightly curled sepals and petals are the source of the name for *H. revoluta*

Ridges such as those on the lip of *H. revoluta* help identify the genus

Hexalectris revoluta grows from coral-like masses

Maturing seed capsules of *H. revoluta*

Hexalectris spicata var. *arizonica* grows in oak forests and oak-lined canyons

The typical flower of *H. spicata* var. *arizonica* never fully opens before self-pollinating

Only rarely does a flower of *H. spicata* var. *arizonica* open fully

Maturing seed capsules of *H. spicata* var. *arizonica*

Plate 13

Hexalectris spicata var. *spicata* in Santa Cruz County, Arizona

Hexalectris warnockii in mixed oak forest in Cochise County, Arizona

It is difficult to separate *H. spicata* var. *spicata* from the rare open flower of *H. spicata* var. *arizonica*

A rare seed capsule of *H. warnockii*

The purple flowers with yellow on the lip help identify *H. warnockii*

Listera convallarioides grows on damp stream banks and in seeps

L. convallarioides blooming in Pima County, Arizona

The narrow claw on the lip is a clue to the identity of *L. convallarioides*

L. cordata can be identified by the deep lobes of the lip

Listera cordata, the heart-leaved twayblade, gets its common name from the shape of its leaves

Plate 15

Malaxis abieticola grows in the same habitat as *M. porphyrea*

A group of *M. abieticola* in Cochise County, Arizona

The green stripes on the lip of *M. abieticola* are clues to its identification

The flowers of *M. abieticola* face in many directions

Seed capsules of *M. abieticola*

The flower spike of *M. corymbosa* is visible in the newly emerged leaf

Malaxis corymbosa grows in damp mossy areas

The circular head of flowers on *M. corymbosa*

Close-up of *M. corymbosa*

Seed capsules of *M. corymbosa*

Plate 17

This flower may simply be a variation within *M. corymbosa* or may represent a new taxon

A slightly different plant, perhaps a variety of *Malaxis corymbosa*

Malaxis porphyrea grows in slightly damper spots in the forest

Each plant of *M. porphyrea* may bear 100 or more purple flowers

A magnifying glass is required to appreciate the flowers of *M. porphyrea*

Plate 18

Malaxis soulei grows in dry places in the forest

A tightly packed flower spike is typical of *M. soulei*

Very few flowers on *Malaxis porphyrea* set fruit

Some plants of *M. soulei* have two-toned green flowers

Most plants of *M. soulei* have monochrome flowers

The leaves on *Piperia unalascensis* are usually faded by the time the flowers open (photographed in Washington)

Close-up of *P. unalascensis* (photographed in Washington)

P. unalascensis is usually laxly flowered (photographed in Washington)

Plate 20

Platanthera aquilonis grows in wet places in northern New Mexico

A flowering stem of *P. aquilonis*

The rhombic lip helps identify *P. aquilonis*

Lip removed to show self-pollination of *P. aquilonis*

Plate 21

The abbreviated leaves are clues to the identity of *Platanthera brevifolia*

P. brevifolia replaces *P. sparsiflora* in southern New Mexico

Side view of *P. brevifolia* showing the long spur

Platanthera huronensis filling a wet meadow in Santa Fe County, New Mexico

Typical greenish white form of *P. huronensis*

White form of *P. huronensis* probably responsible for many of the reports of *P. dilatata* in New Mexico

Densely packed spike of *Platanthera huronensis*

The greenish white color is diagnostic for *P. huronensis* in New Mexico

Platanthera limosa in a wet spot in Pima County, Arizona

A densely flowered spike of *P. limosa*

The yellow-green lip and large column help identify *P. limosa*

P. limosa has a spur much longer than the lip

Platanthera purpurascens in dry habitat in Greenlee County, Arizona

P. purpurascens in wet habitat in Graham County, Arizona

Dense spike of *P. purpurascens*

The slightly dilated lip and small column of *P. purpurascens*

The lip of *P. purpurascens* caught in the petals aids pollination and also exposes the clavate spur

Roots of *P. purpurascens*

Plate 24

Platanthera sparsiflora in Apache County, Arizona

P. sparsiflora var. *ensifolia* in a small wet area in Apache County, Arizona

The sparsely flowered spike of *P. sparsiflora* gives the species its common name

The large column and slightly elliptic lip of *P. sparsiflora* var. *ensifolia*

The linear lip, large column, and spur the same length as the lip are characteristic of *P. sparsiflora*

The spur on *P. sparsiflora* var. *ensifolia* is about equal in length to the lip

Platanthera zothecina streamside in Coconino County, Arizona

The broadly elliptic mostly basal leaves help identify *P. zothecina*

Close-up of *P. zothecina* showing large column and linear elliptic lip

The spur on *P. zothecina* is much longer than the lip

The anther sacs on *P. zothecina* are parallel

Green and normal color forms of *Schiedeella arizonica* growing together in Cochise County, Arizona

An ambush spider awaits a pollinator on a stem of *S. arizonica*

The tiny flowers of *S. arizonica* are best seen with a magnifying glass

The leaves of *S. arizonica* appear several weeks after flowering is over

Plate 27

Close-up of *Schiedeella arizonica*

The red spot on the lip of *S. arizonica* is best viewed from underneath

Seed capsules of *S. arizonica*

Cormlike rhizome of *S. arizonica* with shoot that will be next year's plant

Spiranthes delitescens growing on the damp edge of a cienega in Southeast Arizona

The flower spikes of *S. delitescens* barely reach above the leaves of surrounding grasses

Close-up of *S. delitescens*

Seed capsules of *S. delitescens*

Plate 29

Spiranthes magnicamporum in damp soil above a riverbank in northern New Mexico

The spreading sepals and petals help identify *S. magnicamporum*

The spiral nature of the inflorescence on *S. magnicamporum*

Seed capsules on *S. magnicamporum*

Habitat for *Spiranthes romanzoffiana* in Apache County, Arizona

S. romanzoffiana in a damp meadow in Graham County, Arizona

A form of *S. romanzoffiana* with white floral bracts

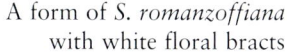

S. romanzoffiana is usually in dense spirals

The sepals and petals form a tight hood over the lip of *S. romanzoffiana*

Seed capsules of *S. romanzoffiana*

Plate 31

The alligator juniper habitat of *Stenorrhynchos michuacanum*

Early in the monsoon season the growth of *S. michuacanum* is lush and green

When *S. michuacanum* flowers at the end of the summer rainy season, the blooming stalks are difficult to see among the drying grasses

Most flowers on a spike of *S. michuacanum* face the same direction

Close-up of *S. michuacanum*

Seed capsules of *S. michuacanum*

Plate 32

Introduction

The orchids, Orchidaceae, comprise perhaps the largest of all plant families, with estimates for the number of naturally occurring taxa running as high as 35,000. With the exception of Antarctica, wild orchids occur throughout the world, even within the Arctic Circle. Over 200 species of orchids are native to the United States, and a surprisingly diverse orchid flora exists within Arizona and New Mexico. In fact, Arizona has 26 native orchid species in 13 genera, and New Mexico has 28 species in 13 genera! The combined orchid flora consists of 35 species in 14 genera. Most species are found in both states, but Arizona has 7 species and 1 genus not seen in New Mexico, and New Mexico has 8 species and 1 genus not seen in Arizona.

Structural characteristics determine whether or not a flower is an orchid, and the botany of orchids involves some specialized terms. A brief description of the orchid flower (Figure 1) is presented here, to introduce the terminology and help the reader understand the discussions that follow. Orchid flowers have three sepals and three petals; two of the petals usually look like and have the same color as the sepals. The third petal, called the *lip*, is usually much larger than the other two, with a different shape and typically a different color. The lip is central to the reproductive cycle of the orchid; it serves as the landing platform for the pollinator. The presentation of the lip results in a flower with bilateral symmetry instead of radial symmetry. An imaginary line drawn from the tip of the uppermost sepal to the center in the bottom of the lip divides the flower into mirror-image halves. The reproductive parts of the flower, the pistils and stamens, are usually separate in most plant families. In

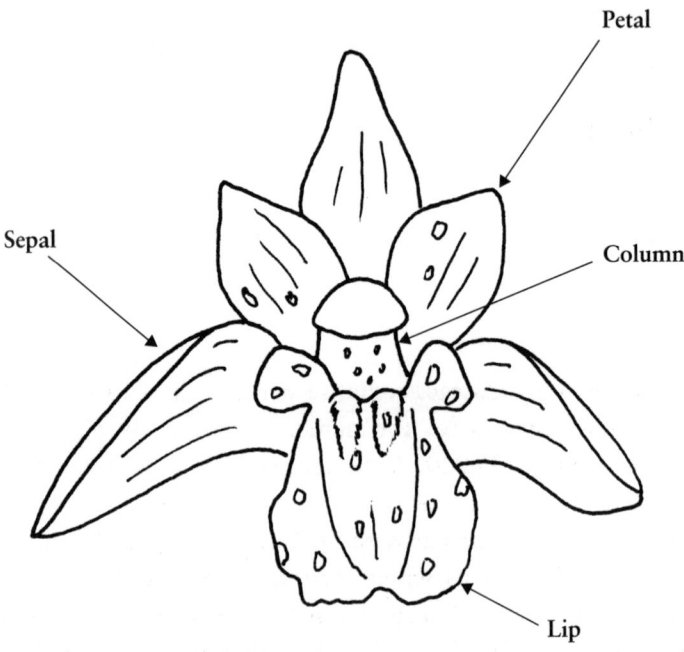

Figure 1. Structure of the orchid flower

the Orchidaceae they are fused into a single structure called a *column*. Orchid pollen usually is removed en masse by the pollinator, and orchid seeds are many and minute. These features are not unique to orchids but are scattered widely in nature. However, if a single flower has all of these features, it is an orchid! Part of the beauty and intrigue of orchids is the many different ways they present the parts that distinguish them, and the role each part plays in pollination. Interested readers are referred to the books by van der Pijl and Dodson (1966) and Dressler (1993) for detailed discussions of orchid biology.

The southwestern states of Arizona and New Mexico are known for their spectacular beauty and variety of landscapes. The Grand Canyon and Carlsbad Caverns are here, as are the northern end of the Sonoran Desert and the southern terminus of the Rocky Mountains. Less known, but equally impressive is the floral diversity of the Southwest. This region

Table 1. Orchids at Range Limit in Either Arizona or New Mexico

Range Limit	Species
Northern	*Hexalectris revoluta, H. warnockii* Malaxis corymbosa, *M. porphyrea, M. soulei, M. abieticola, Platanthera limosa, P. brevifolia, Schiedeella arizonica, Stenorrhynchos michuacanum*
Southern	*Calypso bulbosa, Corallorhiza trifida, Goodyera repens, Listera convallarioides, L. cordata, Platanthera aquilonis, P. huronensis, P. purpurascens, P. zothecina, Spiranthes magnicamporum, S. romanzoffiana*
Western	*Coeloglossum viride, Corallorhiza wisteriana, Hexalectris spicata* var. *spicata, H. spicata* var. *arizonica*
Southwestern	*Cypripedium parviflorum*
Southeastern	*Piperia unalascensis*
Northwestern	*Hexalectris nitida*
Endemic to Arizona	*Spiranthes delitescens*

interests students of native North American orchids because of its confluence of habitats and its rare and unusual plants. Think of Arizona and New Mexico as a great floral crossroad, with major influences converging from the north and south. The northern influence comes via the Rocky Mountains. Even though the Rocky Mountains end in New Mexico, their floral signature extends into Arizona. A distribution map of some of the native orchids would show them in the extreme northern parts of the United States and in Canada, ranging south along the Rocky Mountains, and fanning out from the southern terminus of the Rocky Mountains into adjacent parts of New Mexico and Arizona.

Equally important is the proximity of Mexico, with the resultant abundance of plants more typical of the Mexican Sierra Madres and other parts of Mexico. Because part of the regional flora is common with adjacent Mexico, there are several orchids that occur in Arizona, New Mexico, and a small corner of Texas but nowhere else in the United States. Although the major influences on the southwestern orchid flora are from the north and south, to a lesser extent the California floristic province and the eastern United States are represented here.

Because Arizona and New Mexico represent somewhat of an orchid melting pot, 29, or an amazing 83 percent, of the 35 native orchids, including the one endemic, are at a limit of their range, as shown in Table 1. Ten are at their northern limit; 11, at their southern limit; and 4, at their western limit. One orchid is at its southwestern limit; 1, at its southeastern limit; and 1, at its northwestern limit. Perhaps because they are at the limits of their range, many species are relatively rare here, though they may be more plentiful elsewhere.

Habitat

All of the southwestern orchids, and the vast majority of temperate zone orchids, are terrestrial (grow in the ground), in contrast to most species in the family, which are either epiphytic (grow attached to trees) or lithophytic (grow on rocks). Most of these orchids occur above elevations of 5000 feet (1524 meters), but *Epipactis gigantea* is found as low as 1200 feet (366 meters) in the Southwest and much lower elsewhere. Most orchids in this area are found in the mountains; a single mountain range may contain 10 species. Individuals of all species of wild orchids in Arizona and New Mexico grow within sight of a prepared trail or road in at least part of their range. Wild orchids grow in riparian areas and wetlands, juniper woodland, mixed oak woodland, and mixed and coniferous forests. These four loosely defined habitats are discussed below, with notes on the orchids that may be expected in each.

Diverse riparian and wetland habitats are home to 13 species of orchids including *Epipactis gigantea*, several *Platanthera* species, *Listera convallarioides*, *Malaxis corymbosa*, and *Spiranthes* species. The orchids grow in damp soil, often directly on stream banks, though just as often they are scattered away from the surface water but in permanently damp soils. A unique habitat in the Southwest is freshwater wetlands nestled in semiarid grasslands. These areas, known as *cienegas*, are usually fed by springs. Some are saucer-shaped, with the floor of the cienega saturated or covered with water. Parts of the slopes seasonally run with water, with dampness decreasing up the slope until it merges with the drier grassland.

A wetland or riparian type of habitat known as a *seep* manifests itself on the sides of canyons, valleys, or river-cut gorges. A seep occurs when an impervious rock layer is exposed to the surface by erosion, geologic action, or a river or stream channel. Groundwater percolating through the soil reaches the impervious layer and follows it to the surface. If the water flow is continuous and measurable, the result may be called a *spring*, but if the area is just constantly damp or dripping, it is usually called a *seep*. If high on the canyon side, the seep may be called a *hanging garden*. Spectacular examples of seeps can be seen in the canyonlands in the north, or in the Grand Canyon drainage, but they occur in widely scattered locations throughout both states.

At least 4 species of juniper are found throughout the Southwest, but alligator juniper (*Juniperus deppeana*) dominates, particularly in the southern parts of these states. Near the foothills the trees are dense enough to properly be called *woodlands*. At slightly higher elevations, and deeper in the canyons, the juniper woodlands are replaced by mixed oak woodlands consisting of oaks, junipers, and various pines. These two woodland habitats are home to 6 unusual and rare orchids. *Stenorrhynchos michuacanum* grows in the alligator juniper woodlands in the southeastern corner of Arizona, and 5 *Hexalectris* species grow in mixed oak woodlands in the southern parts of both states.

The fourth habitat is mixed and coniferous forests, and 17 species of orchids are found therein. Mixed and coniferous forests occur from about 7000 to above 10,000 feet (2134 to 3048 meters) and consist of mixed and pure stands of aspen, pine, fir, and spruce, with varying amounts of oak and juniper. Within this large range in elevation is a broad mix of microhabitats, from mesic to xeric. The orchids to be expected within the forest vary with moisture content. In the dampest region, various ferns and mosses are thick and may cover the forest floor. This is the area to look for *Calypso bulbosa, Goodyera repens, Listera cordata, Corallorhiza trifida*, and *Cypripedium parviflorum*. In slightly drier areas, *Corallorhiza maculata, C. striata, C. wisteriana, Malaxis abieticola, M. porphyrea, Goodyera oblongifolia, Coeloglossum viride, Epipactis helleborine*, and *Schiedeella arizonica* can be found. The more xeric areas will host *Malaxis soulei*.

The boundaries between these habitats are not always clear. In any event, the orchids pay scant attention to the boundaries scientists create for ease of classification. In the mixed and coniferous forests, for example, both *Corallorhiza trifida* and *C. maculata* grow equally as well in damp, mossy habitats with plenty of competing herbaceous undergrowth as in much drier open forest. Both *Stenorrhynchos michuacanum* and *Malaxis soulei* grow in alligator juniper woodlands and in mixed and coniferous forests. *Hexalectris spicata* var. *spicata* and *H. spicata* var. *arizonica* grow equally well in mixed oak woodlands as in the lower reaches of mixed and coniferous forests.

Distribution

Wild orchids are widely distributed in Arizona, appearing in 12 of the state's 15 counties. Table 2 lists the number of orchid species that occur in each county. These statistics, however, are a bit misleading. Since Arizona has relatively few counties, several of them are large, and proper orchid habitat exists in only a portion of their total landmass. Instead of being uniformly spread within a county, the orchids are found mainly in the areas where total precipitation and elevation, two factors that determine the dominant plant communities, are conducive to their growth.

With regard to orchid communities, Arizona can be divided into four main regions. The least interesting in terms of orchids (though it has a unique beauty and flora) is the southern desert, composed of major portions of western Pima County, most of Pinal and Maricopa Counties, parts of Yavapai and Mohave Counties, and all of La Paz and Yuma

Table 2. Species Distribution by County in Arizona

County	No. of species	County	No. of species
Apache	16	Mohave	1
Cochise	17	Navajo	6
Coconino	13	Pima	13
Gila	7	Pinal	0
Graham	10	Santa Cruz	12
Greenlee	12	Yavapai	2
La Paz	0	Yuma	0
Maricopa	1		

Counties. Rainfall here is usually less than 8 inches (20.24 cm) per year. This region is probably devoid of orchids, although *Epipactis gigantea* can never be ruled out and may be growing at some hidden spring in an isolated canyon.

The largest region encompasses the northern third of Arizona and is part of the Colorado Plateau. The Colorado Plateau extends into Utah, Colorado, and New Mexico and includes the Colorado River drainage and the Grand Canyon. Within the Colorado Plateau are large portions of Apache, Navajo, Coconino, and Mohave Counties and the subregion north of the Grand Canyon known as the Kaibab Plateau. Within Arizona, the western portion of the Colorado Plateau is mostly canyonlands with peak elevations between 5000 and 7000 feet (1524 and 2134 meters), although interspersed mountain ranges reach higher elevations. Yearly precipitation is 12 inches (25 cm) or less. Wet spots in canyons harbor a few orchids such as *E. gigantea* and some *Platanthera* species. Some of the higher ranges such as the Lukachukai Mountains have some *Corallorhiza* and *Goodyera* species.

The Kaibab Plateau north of the Grand Canyon in Coconino and Mohave Counties, technically part of the Colorado Plateau, is dramatically different in habitat. Though largely flat, it is much higher in elevation, mostly between 8000 and 9000 feet (2438 and 2743 meters). The region near the North Rim of the Grand Canyon in Coconino County is generally wetter than the eastern Colorado Plateau, with total yearly precipitation above 20 inches (about 50 cm) at higher elevations. Consequently, the vegetative forms change, and part of the Kaibab Plateau is covered with coniferous forest. Orchids here include *Calypso bulbosa, Corallorhiza maculata, C. striata,* and *Spiranthes romanzoffiana.*

Most orchids in Arizona, both in total number of plants and in diversity of species, occur in the remaining two regions, the Mogollon Rim and the Sky Islands. The Mogollon Rim, a zone of sometimes precipitous uplift, stretches from western New Mexico on a northwest axis through central Arizona, ending near Flagstaff. The rim and associated mountain ranges stretch across parts of Greenlee, Graham, Apache, Navajo, Gila, Coconino, and Yavapai Counties. There are few roads across the face of the rim, and except for Interstate Highway I-17 they are steep and twisting in places. This area has the largest amount of precipitation in Arizona, more than 16 inches (40 cm) per year and in some locales over 36 inches (90 cm) per year. *Calypso bulbosa* grows here, as do 3 *Corallorhiza* species, several *Platanthera* species, *Schiedeella arizonica, Goodyera oblongifolia,* and 2 *Malaxis* species. A few of Arizona's wild orchids, such as *Cypripedium parviflorum, Goodyera repens,* and *Coeloglossum viride,* grow only in the area of the Mogollon Rim.

The Sky Islands, the fourth orchid region in Arizona, get their name from the high mountain ranges that stand above the expansive desert grasslands like islands in the ocean. In less than an hour it is possible to drive from Sonoran Desert habitat populated with barrel cactus, ocotillo, and saguaros, to forests of aspen, fir, pine, and permanent streams, passing through juniper and oak woodlands on the way. The Sky Islands are in the extreme southeastern corner of Arizona, in the counties of Cochise, Pima, Graham, and Santa Cruz. Precipitation is about the same as in the Mogollon Rim area, with mountain peaks getting over 36 inches (90 cm) per year. The orchid flora in the Sky Islands is fascinating. Some northern influence is still evident in such species as *Goodyera oblongifolia, Spiranthes romanzoffiana, Corallorhiza maculata, C. striata,* and *C. wisteriana.* However, the southern flora is nearly dominant, with many orchids more common to Mexico such as *Malaxis* and *Hexalectris* species, *Platanthera limosa,* and *Stenorrhynchos michuacanum.*

The orchid distribution in New Mexico is limited even more geographically than it is in Arizona. New Mexico has more and smaller counties than Arizona, 33 compared to 15, and correspondingly more (10) without any record of wild orchids. The entire southern region of the state adjacent to either Mexico or Texas is Chihuahuan Desert and Chihuahuan Desert grasslands. As in the desert regions of Arizona, total annual precipitation in the deserts of New Mexico is less than 12 inches (25 cm). Except for in the Sky Island ranges of the Sacramento Mountains in Otero County and the Guadalupe Mountains in Eddy County, there have been no orchids reported in southern New Mexico. The

Table 3. Species Distribution by County in New Mexico

County	No. of species	County	No. of species
Bernalillo	6	McKinley	1
Catron	13	Mora	5
Chaves	0	Otero	12
Cibola	3	Quay	0
Colfax	6	Rio Arriba	12
Curry	0	Roosevelt	0
De Baca	0	San Juan	3
Dona Ana	0	San Miguel	13
Eddy	2	Sandoval	12
Grant	12	Santa Fe	9
Guadalupe	0	Sierra	8
Harding	1	Socorro	3
Hidalgo	0	Taos	11
Lea	0	Torrance	3
Lincoln	10	Union	1
Los Alamos	7	Valencia	3
Luna	0		

orchid flora is also sparse in the eastern third of the state, which is dominated by high plains and grasslands. Precipitation is higher here, up to 20 inches (50 cm), but probably because of the lower elevation, orchids are few. Seven of the 9 counties along or close to the border with either Texas or Oklahoma have no reports of orchids, while Union County has only *Corallorhiza maculata* and Harding County has only *Epipactis gigantea*.

The northwestern corner of New Mexico, composed of all or portions of San Juan, Rio Arriba, Sandoval, McKinley, and Cibola Counties, is part of the Colorado Plateau and is mostly without orchids. The exceptions are on the few higher mountain ranges in the region. Precipitation varies from under 12 to over 25 inches (30 to 62 cm) in the mountains. The Chuska Mountains on Navajo Nation lands in San Juan County are high enough to host *Corallorhiza maculata* and *Cypripedium parviflorum*. The presence of these two species strongly suggests that the orchid flora in the Chuska Mountains is much more varied but perhaps underreported, since many other orchids are associated with these two species in other parts of their range.

The west-central part of New Mexico, including Catron, Socorro, Grant, and Sierra Counties, contains the eastern terminus of the Mogollon Rim. The orchid flora in the rim region in New Mexico is similar to that in Arizona, with *Calypso bulbosa*, *Cypripedium parviflorum*, several *Platanthera* species including *P. brevifolia*, *Schiedeella arizonica*, *Hexalectris spicata*, and 3 *Corallorhiza*, 2 *Goodyera*, and 3 *Malaxis* species. Precipitation is equal to that in the Mogollon Rim area of Arizona, over 36 inches (90 cm) per year in the mountains.

The north-central region of New Mexico contains the southern end of the Rocky Mountains in Rio Arriba, Taos, Colfax, Mora, San Miguel, Santa Fe, Sandoval, and Los Alamos Counties. Because the Rocky Mountains are a bridge to more northern floristic provinces, the orchids are those found in Colorado or at even higher latitudes. *Calypso bulbosa* and *Cypripedium parviflorum* are plentiful, and *Platanthera huronensis* and *P. aquilonis* grow in only this region of Arizona and New Mexico. Two other orchid species typical of higher latitudes are *Corallorhiza trifida* and *Listera cordata*, though both are relatively rare here. The increase in orchid diversity is predicted by the precipitation, which is over 25 inches (62 cm) per year in much of the region and up to 36 inches (90 cm) in the mountains.

Another way of looking at distribution is to study the number of counties in which each species occurs, as shown in Table 4. The two states have a combined total of 48 counties. The data suggest that *Corallorhiza* is the most widely distributed genus and *C. maculata* the most widely distributed species, occurring in 28 counties. *Corallorhiza striata* is the second most widely distributed species and *C. wisteriana* the fourth. The species with the lowest frequency of occurrence are *Epipactis helleborine*, *Hexalectris nitida*, *H. warnockii*, and *Piperia unalascensis*. Each of these occurs in only 1 county. These data relate only the relative frequency of occurrence in Arizona and New Mexico and do not imply rareness in an absolute sense. *Epipactis helleborine* is invasive and spreading throughout the United States and Canada. *Piperia unalascensis* is widespread at higher elevations in the western part of the United States. *Hexalectris nitida* and *H. warnockii*, however, are rare not only in these two states but also throughout their range.

Table 4. Number of Counties in Which Each Species Occurs

Species	Arizona (maximum 15)	New Mexico (maximum 33)
Calypso bulbosa var. *americana*	3	11
Coeloglossum viride var. *virescens*	1	4
Corallorhiza maculata	8	20
Corallorhiza striata	8	13
Corallorhiza trifida	0	3
Corallorhiza wisteriana	9	7
Cypripedium parviflorum var. *pubescens*	3	8
Epipactis gigantea	11	7
Epipactis helleborine	0	1
Goodyera oblongifolia	6	11
Goodyera repens	2	7
Hexalectris nitida	0	1
Hexalectris revoluta	3	0
Hexalectris spicata var. *arizonica*	3	1
Hexalectris spicata var. *spicata*	3	1
Hexalectris warnockii	1	0
Listera convallarioides	3	0
Listera cordata	0	4
Malaxis abieticola	2	3
Malaxis corymbosa	2	0
Malaxis porphyrea	5	7
Malaxis soulei	9	7
Piperia unalascensis	0	1
Platanthera aquilonis	0	5
Platanthera brevifolia	0	5
Platanthera huronensis	0	3
Platanthera limosa	3	1
Platanthera purpurascens	4	7
Platanthera sparsiflora	4	4
Platanthera zothecina	3	0
Schiedeella arizonica	6	5
Spiranthes delitescens	2	0
Spiranthes magnicamporum	0	3
Spiranthes romanzoffiana	3	6
Stenorrhynchos michuacanum	3	0

Blooming Seasons

Because of the range of elevations and relatively mild weather, the orchid blooming season in Arizona and New Mexico extends for nearly 7 months, from the end of March to the end of October. The season

Table 5. Blooming Seasons

Species	J	F	M	A	M	J	J	A	S	O	N	D
Epipactis gigantea					▬▬▬▬▬▬▬							
Corallorhiza wisteriana					▬▬▬▬▬▬▬							
Corallorhiza striata						▬▬▬▬▬						
Schiedeella arizonica						▬▬						
Calypso bulbosa var. americana						▬▬▬						
Platanthera sparsiflora						▬▬▬▬▬						
Corallorhiza maculata						▬▬▬▬						
Hexalectris revoluta						▬						
Cypripedium parviflorum var. pubescens						▬▬						
Hexalectris spicata var. spicata							▬					
Corallorhiza trifida							▬▬					
Platanthera aquilonis							▬▬					
Platanthera zothecina							▬▬					
Listera cordata							▬▬					
Epipactis helleborine							▬▬					
Platanthera purpurascens							▬▬▬▬▬					
Coeloglossum viride var. *virescens*							▬▬					
Platanthera huronensis							▬▬▬					
Hexalectris nitida							▬▬					
Piperia unalascensis							▬▬▬					
Platanthera brevifolia							▬▬▬▬▬▬▬					
Platanthera limosa							▬▬▬▬					
Goodyera oblongifolia							▬▬▬▬▬					
Malaxis corymbosa							▬▬▬▬▬					
Listera convallarioides							▬▬▬▬▬					
Spiranthes delitescens							▬					
Malaxis abieticola							▬▬▬▬					
Malaxis porphyrea							▬▬▬▬					
Malaxis soulei							▬▬▬▬					
Hexalectris spicata var. arizonica							▬▬					
Goodyera repens							▬▬					
Spiranthes romanzoffiana							▬▬▬▬▬					
Hexalectris warnockii							▬▬					
Spiranthes magnicamporum									▬			
Stenorrhynchos michuacanum										▬		

starts with *Epipactis gigantea* and ends with *Stenorrhynchos michuacanum*. The flowering period for each species is shown in Table 5. Entries in the table are based on field and herbarium research and represent the flowering season in these two states. Flowering may start and end at different times in other states. For *Epipactis helleborine, Hexalectris nitida, Listera cordata*, and *Piperia unalascensis*, the blooming season is estimated because there are only one or two records of the plants in this region.

The start and end of each flowering period shown in Table 5 are the extremes of the historical record and not expected dates each year. The start or end of the season for any species can vary by more than 2 weeks, depending on seasonal temperatures and precipitation. Because the temperature is cooler at higher elevations, the season is delayed and ends correspondingly later than for the same species at lower elevations. For example, *Calypso bulbosa* might bloom as early as 15 May at elevations of 8000 to 9000 feet (2438 to 2743 meters) in some years, but it is more likely to begin blooming in late May and usually finishes blooming by the middle of June. However, if the searcher is willing to hike 8 to 10 miles (15 to 17 km) with an elevation gain of over 1000 feet (305 meters), *Calypso* sometimes can be found in bloom after 4 July.

Though not readily apparent from Table 5, there are two overlapping growing seasons for orchids in the Southwest. The primary season starts with *Epipactis gigantea* in March and continues through to September when *Spiranthes magnicamporum* flowers. The secondary bloom period is timed to a seasonal wind shift that brings moist air north from Mexico. This moist air flow results in summer thunderstorms, referred to as the *monsoon season*. About half of the area's annual rainfall comes during the 2 months of the summer monsoon season. Because of the monsoon rains, several orchids more typical of Mexico grow in this region. Most of these monsoon orchids are known in the United States only from a few isolated spots in Arizona, New Mexico, and Texas. They are most common near the border with Mexico, and in particular in Cochise, Graham, Pima, and Santa Cruz counties in southeastern Arizona. These orchids typically do not even appear above ground until after the summer rains begin. The monsoon orchids are *Malaxis corymbosa, M. por-*

phyrea, M. soulei, M. abieticola, Hexalectris warnockii, and *Stenorrhynchos michuacaum.* Depending on the onset of the summer rains, they start blooming in early to mid-July and may remain in bloom as late as the end of October.

The best months for general orchid hunting can be gleaned from Table 5. Only 1 species blooms in March and only 2 in April. At the other end of the season, only *Stenorrhynchos michuacanum* blooms in October. The peak months for flowering are June, July, and August, with 17, 27, and 16 species, respectively, in bloom.

Conservation

Many of the orchids native to the Southwest are rare in at least a portion of their range, and both states provide some measure of official protection and monitoring. Within Arizona, orchids are protected by the Native Plant Law and Antiquities Act. All of the orchids in Arizona are classified as Highly Safeguarded Protected Native Plants, which is the highest level of protection under the law. The plants, fruits, and seeds are equally protected, and permits are needed even for scientific collecting. In New Mexico legal protection is provided to only those plants on List 1 (L1), the highest of four classifications.

The Heritage Data Management System of the Arizona Game and Fish Department (1999) maintains a database of all the rare plants in the state. The database tracks six factors for each listed plant: number of occurrences, estimated population size, estimated state range, trends in distribution in the state, estimated number of protected occurrences, and degree to which the plant is threatened. The six factors are considered when a rank is assigned to each species. The assigned ranking is based on the Nature Conservancy's G/S system, wherein each species is assigned both a global (G) and state (S) ranking. Each plant can be assigned a G or S ranking from 1 to 5, with the implications shown in Table 6.

Generally the Heritage Data Management System does not track plants with state rankings above S4, so while all orchids are protected

Table 6. Nature Conservancy Ranking System

G	S	Ranking
G1	S1	Very rare: 1 to 5 occurrences
G2	S2	Rare: 6 to 20 occurrences
G3	S3	Uncommon or restricted: 21 to 100 occurrences
G4	S4	Apparently secure: more than 100 occurrences, though it could be quite rare in parts of its range
G5	S5	Demonstrably secure: more than 100 occurrences

Table 7. Species Ranking for Arizona

Species	Ranking	Species	Ranking
Calypso bulbosa	G5/S4	*Listera convallarioides*	G5/S1
Cypripedium parviflorum var. *pubescens*	G4/S1	*Malaxis corymbosa*	G4/S3S4
Goodyera repens	G5/S3	*Malaxis porphyrea*	G2G3/S2
Platanthera limosa	G4/S3	*Malaxis abieticola*	G4/S1
Coeloglossum viride	G5/S1	*Schiedeella arizonica* (as *Schiedeella parasitica*)	G4/S4
Platanthera zothecina	G2/S1	*Spiranthes delitescens*	G1/S1
Hexalectris spicata	G5/S3S4	*Spiranthes romanzoffiana*	G5/S3S4
Hexalectris warnockii	G2/S1	*Stenorrhynchos michuacanum*	G5/S3

in Arizona, not all are being actively tracked. The rankings for the orchids being tracked are shown in Table 7.

Hexalectris revoluta undoubtedly will be added to the list in the near future, but when state rankings were last revised, it had not yet been published as part of the Arizona flora.

The Forestry Division of the Energy, Minerals and Natural Resources Department maintains the Inventory of Rare and Endangered Plants of New Mexico (Sivinski and Lightfoot 1995). The third edition is available on the Internet via New Mexico's official website. Plants are ranked by list, with List 1 being plant species endangered in New Mexico. List 1A is for plants listed as threatened or endangered under the Federal Endangered Species Act. List 1B is for plants so rare within the state that unregulated collection could jeopardize their survival in New Mexico. Plants on either List 1A or List 1B are protected by law, but plants on the other lists are not afforded protection. List 2 is for plants considered

Table 8. New Mexico's Rare Orchids

Species	List	Species	List
Cypripedium parviflorum var. pubescens (as C. pubescens)	L1	Malaxis abieticola	L3
Epipactis gigantea	L2	Piperia unalascensis	L3
Hexalectris nitida	L1	Platanthera dilatata	L2
Hexalectris spicata	L1	Spiranthes magnicamporum	L1

rare and sensitive in New Mexico because of restricted distribution or low numerical density. List 3 is the review list and contains taxa about which more knowledge is required before they can be considered for either List 1 or List 2. The listed orchids are shown in Table 8.

Platanthera dilatata most likely does not occur in New Mexico and should be removed from the list. *Corallorhiza trifida* should be added to List 1A, and *Listera cordata* should be added to List 2.

Key to the Orchid Genera in Arizona and New Mexico

After a flower has been identified as an orchid, the natural next question is, "What genus is it?" The key below should help answer that question. Like all artificial keys, it is imperfect, but it should lead to a correct determination of genus in most cases. It is constructed of two contrasting statements at the same level of indentation (e.g., 1 and 1a). If the first statement (e.g., 1) applies, follow the contrasting indented statements below it. If it does not apply, continue down the key until reaching the second contrasting statement indented (e.g., 1a). For example, suppose you wanted to determine the genus of an orchid with many tubular, spurless, white flowers, spiraled tightly about the inflorescence axis in three rows, and with linear leaves clustered mostly at the base. The first statement "1. Flowers one or two" does not apply, so skip down to "1a. Flowers more than two." It does not have a spur, so skip statement 3 and go down to "3a. Flowers without a spur." The flowers are tubular (statement 6), so look at the next level "7. Flowers spiraled about the inflorescence axis," which also matches. Congratulations! You have determined that the flower is in the genus *Spiranthes*! The information that the flowers were white or that the leaves were linear was not needed. To determine the species, consult the key supplied with the genus treatment.

Key to Orchid Genera

1. Flowers one or two
    2. Lip a pink to white pouch with yellow hairs and markings at pouch opening   *Calypso*
    2a. Lip a yellow pouch with or without reddish dots at pouch opening   *Cypripedium*
1a. Flowers more than two
    3. Flowers with a spur from small and saccate to long and slender
        4. Leaves still fresh and green at flowering
            5. Apex of lip acute   *Platanthera*
            5a. Apex of lip two- or minutely three-lobed   *Coeloglossum*
        4a. Leaves faded or yellowing at flowering   *Piperia*
    3a. Flowers without a spur
        6. Flowers tubular
            7. Flowers spiraled about inflorescence axis   *Spiranthes*
            7a. Flowers not spiraled about inflorescence axis
                8. Center of lip with red or orange dot   *Schiedeella*
                8a. Center of lip without red or orange dot   *Stenorrhynchos*
        6a. Flowers not tubular
            9. Leaves not present at flowering
                10. More than two raised ridges along entire length of central lobe of lip   *Hexalectris*
                10a. Two or fewer ridges along less than upper half of central lobe of lip   *Corallorhiza*
            9a. Leaves present at flowering
                11. Leaves in basal rosette   *Goodyera*
                11a. Leaves not in basal rosette
                    12. Leaves one   *Malaxis*
                    12a. Leaves more than one
                        13. Leaves two   *Listera*
                        13a. Leaves more than two   *Epipactis*

Calypso Salisbury

Paradisus Londinensis: Pl. 89. 1807.
Etymology: *Calypso* is from the Greek word for the sea nymph of Homer's *Odyssey*.

Calypso (ka-lip'-soe) is a monotypic, circumpolar genus with four recognized varieties: *Calypso bulbosa* var. *bulbosa* in Europe and Asia; *C. bulbosa* var. *speciosa* in Japan; and two varieties, *C. bulbosa* var. *americana* and *C. bulbosa* var. *occidentalis*, in North America. *Calypso* is one of only three summer deciduous (without summer leaves) North American orchid genera. The other two genera are *Tipularia* and *Aplectrum*, neither of which occurs in this region. Dressler (1993) placed *Calypso* and eight other genera in the tribe Calypsoeae. The North American genera in the tribe are *Calypso, Corallorhiza, Tipularia,* and *Aplectrum*. Linnaeus originally included *Calypso* within *Cypripedium* because of its pouch-shaped lip, but Salisbury separated them based on differences in the lip and column. The lip in *Calypso* is covered by a lamina (an apronlike blade over the pouch) whereas the lip in *Cypripedium* is not, and *Cypripedium* has two anthers while *Calypso* has a single anther.

Geographical distribution and morphological characters distinguish the two American varieties of *Calypso bulbosa*. In the past some authorities have argued they are distinct species, but current scholarship accepts recognizing the differences at the varietal level. *Calypso bulbosa* var. *americana*, the more widely distributed race and the one in Arizona and New Mexico, has bright yellow markings and hairs at the opening to the pouch. *Calypso bulbosa* var. *occidentalis* is limited to the far western regions of the United States and Canada. Its throat markings and hairs are white; its hairs are fewer, and it generally has larger leaves and flowers. According to Long (1980), the distributions of the two varieties overlap only in British Columbia.

Calypso bulbosa (Linnaeus) Oakes var. *americana* (R. Brown) Luer

Native Orchids of the United States and Canada: 336. 1975.

Etymology: The specific epithet refers to the bulblike nature of the corms.

Synonymy:
Cypripedium bulbosum Linnaeus, Species Plantarum 2: 951. 1753.
Basionym: *Calypso americana* R. Brown, in Aiton, Hortus Kewensis, ed. 2, 5: 208. 1813.
Calypso borealis (Swartz) Salisbury, Paradisus Londinensis: pl. 89. 1807.
Cytherea borealis (Swartz) Salisbury, Transactions Horticultural Society of London I: 301. 1812.
Calypso bulbosa (Linnaeus) Oakes, in Thompson, History of Vermont I: 200. 1842.
Cytherea bulbosa (Linnaeus) House, Bulletin Torrey Botanical Club 32: 382. 1905.

Common names: calypso, fairy slipper, hider of the north, Venus' slipper, deer's head orchid.

Plate 1

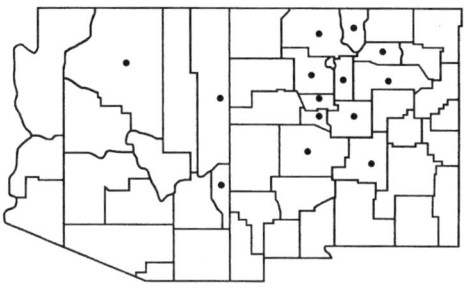

Map 1. Distribution of *Calypso bulbosa* var. *americana*

Description

Plant: typically 10 to 14 cm tall but can be 7 to 21 cm tall.

Roots: leaf and scape rise from ellipsoidal to ovoidal corm with two or three short, slender roots; sometimes coralloid masses develop on corms.

Leaf: solitary, basal, plicate, dark green, ovate, typically 3.5 × 2.5 cm, range from 2.7 × 2 cm to 4.8 × 3 cm; short petiole entirely separate from flower scape.

Scape: purplish with two or three sheathing tubular bracts, glaucous.

Floral bracts: lanceolate, partially sheathing stem, same color as sepals, 0.7 to 1.7 cm long.

Flowers: one, rarely two, purplish to pink to whitish; natural spread of flower between 2 and 3.5 cm.

Sepals and petals: nearly identical, purplish to pink to whitish, lightly veined, lanceolate 1.4 to 2.0 cm long, 0.25 to 0.3 cm wide.

Lip: pouch — (or slipper)-shaped, typically 1.6 × 0.7 cm; top of pouch covered with lamina extending beyond sides of the pouch like an apron; lamina white to purple with purplish spots; area near orifice of pouch is yellow and decorated with three ridges bearing yellow hairs; inside of pouch with purple to reddish veins; bifurcate tip of pouch extends beyond margin of lamina.

Column: shaped like an inverted saucer with wide side lobes, purplish to pink to whitish; extends over the orifice of the pouch; two pairs of yellow pollinia; anther cap arrowhead shaped.

Capsule: ellipsoidal, 1.8 × 0.5 cm; nodding at first, straightening to nearly erect as it matures.

Correll (1950) called *C. bulbosa* (bul-bow'-sa) var. *americana* "the most beautiful terrestrial orchid in North America." It is easy to understand why when viewing this forest gem. The sepals, petals, and bract, all the same color, are held above the colorful lip like a crown. The lip, a complete slipper or pouch, gives rise to the common name of fairy slipper, though *Calypso* is not closely related to the other genus of slipper orchids found in the Southwest, *Cypripedium*. *Calypso* has a distinct, pleasant aroma, which is highly variable in intensity. The aroma is stronger on fresh flowers than on older ones. A typical plant has one flower stem with a single flower. The flower and leaf rise from a shallow cormlike rhizome. A new corm is formed each year, but corms from the last one or two growing seasons remain firm and attached to the new growth. Two flowers on a single scape occur infrequently, but usually both are not prime examples, with one malformed or held at an unnatural angle.

The color of the sepals, petals, and bract varies from the usual purple to shades of pink, and even whitish tones. A white-flowered form, called *C. bulbosa* var. *americana* f. *candida* Whiting and Catling, occurs in parts of its range but has not been documented in either Arizona or New Mexico. Sepals and petals are nearly identical in shape, with the dorsal sepal only slightly longer than the others. Lip color varies from plant to plant. The slipper is cream to purplish, with darker red to purple veins. A yellow blotch and three rows of fine yellow hairs at the orifice of the pouch extend a few millimeters into it. The lamina is mostly white with purple spots, with faint purple infusion around the margins of the pouch. The lamina has five or six dark reddish purple spots in the yellow area near the pouch orifice and additional smaller spots on the edges of the orifice. Sometimes the reddish-purple dots occur in the area of the hairs at the opening to the pouch. The hairs within the dots are then reddish-purple instead of yellow. On many plants, the majority in some locations, the lamina is solid pink to purple, with the usual yellow marking and hairs at the opening to the slipper. Plants with the pink to purple lips have been given the name *C. bulbosa* var. *americana* f. *rosea* P. M. Brown. Plants with many shades and amounts of coloring in the lip are in the same populations, indicating a continuous rather than discrete color variation.

The pollination mechanics of *C. bulbosa* var. *americana* were studied by Mosquin (1970) and Boyden (1982). Both reported that general food mimicry rather than a specific reward attracts pollinators. The yellow hairs at the entrance to the pouch apparently resemble anthers on a food source for the pollinators. However, the bifurcate structures at the end of the pouch that might appear to be nectaries are not, as the flowers contain no nectar. Mosquin identified the pollinators as queen bees of the genera *Pyrobombus*, *Bombus*, and *Psithyrus*. The bee visits the flower in search of food shortly after emerging in the spring. After landing on the lamina of the lip, it attempts to enter the pouch, searching for the nonexistent food. While exiting, the bee makes contact with the column overhanging the opening to the pouch, and pollen is deposited on the bee. Pollen granules are then transferred to the next flower visited. Mosquin and Boyden reported fairly low insect visitation rates. They speculated that the bees are fast learners and after just a few visits realize there is no food in the flowers. Only 10.4 percent of over 1500 flowers studied by Mosquin showed evidence of being visited by bees. Proctor and Harder (1995) documented a much higher pollination rate, up to 25 percent.

Distribution

Calypso bulbosa var. *americana* is widely distributed in North America from the northeastern United States, across Canada, and into Alaska. It occurs in the Rocky Mountain states and in New Mexico and Arizona. Within New Mexico it is in Bernalillo, Lincoln, Mora, Rio Arriba, San Miguel, Sandoval, Santa Fe, Socorro, Taos, Torrance, and Valencia Counties. Because of habitat and its presence in adjacent counties, it probably also occurs in Catron, Colfax, and Grant Counties. Within Arizona, it occurs in Apache, Coconino, and Greenlee Counties.

Habitat

The most common habitats of *C. bulbosa* var. *americana* are either in pure stands of spruce, pine, fir, and aspen or in mixed conifer and aspen forests at elevations between 8000 and 10,800 feet (2440 and 3300 meters). Within those areas it grows in densely forested spots, in open areas within the forest, and on the edges of meadows. Terrain is flat to very steep, with many plants growing in the transition region between the gently sloping and steeper areas. Most often it is found in light to heavy shade, but sometimes it is in nearly direct sunlight. Usually it is rooted shallowly in deep duff or mosses and occasionally grows on top of rotting logs. Often the top of the corm is visible above the duff. On more deeply rooted plants, the leaf appears almost stemless as it barely clears the humus. Calypso may grow on the banks of streams but more often avoids very damp to wet places.

Blooming Season

The blooming season for *C. bulbosa* var. *americana* is usually between late May and late June. In years of heavy snow or late springs, blooming may not start until early June, and at the upper end of its elevation range may continue until early July. Unpollinated flowers remain fresh for about 2 weeks, and buds open over a week or more, so at any elevation plants are in bloom for between 3 and 4 weeks. Pollination causes flowers to fade rapidly. Proctor and Harder (1995) reported that flowers start to fade within 3 days of pollination. Pollinated flowers go through color and shape changes, resulting in flowers with sepals and petals collapsed over the lip, and with a pinkish orange color.

The growth cycle of *C. bulbosa* var. *americana* differs considerably from the cycles of most of the other orchids found in the Southwest. In the fall, as early as September, a newly developed corm sprouts the leaf that lasts through the winter, surviving under the snow in cold areas. Mousley (1924) studied in detail the underground development of *C. bulbosa* var. *americana* and observed that the bud also develops in the fall, usually remaining just underground through the winter. Development of the bud resumes in early spring. The flower is able to withstand late frosts, and blooming plants are often within sight of retreating snow banks. Some bloom within a few weeks of being free of snow. Shortly after flowering, the leaf wilts for the summer, and the capsules dehisce in about 8 weeks.

Several other orchids occur in the same habitat as *C. bulbosa*. *Corallorhiza maculata*, *C. striata*, and *C. wisteriana* are frequent blooming companions as is *Schiedeella arizonica*. Less often, and then only above 9000 feet (2743 meters), *Coeloglossum viride* and *Listera cordata* may co-occur with calypso. Later in the season *Platanthera purpurascens* and *Malaxis soulei* may bloom nearby. Calypso is commonly found within a few centimeters of *Goodyera oblongifolia*, and in the damper portions of its habitat, *G. repens*, although both bloom long after calypso's leaf has withered.

Conservation

Calypso bulbosa var. *americana* is widespread and locally common in parts of Arizona and New Mexico. In Arizona it is ranked G5/S4, which means it is considered demonstrably secure across its entire range and has more than 100 occurrences in the state. It is common enough in New Mexico that it does not have an endangered ranking there. Though not considered rare in these two states, it is not as common in other areas of its range. Chapman (1997) reported decreasing numbers in the northeastern areas of the country in recent years. Throughout its range, it is subject to destruction by logging and development. However, the most serious threat is people collecting calypso for cultivation in gardens.

Notes and Comments

Calypso is a poor candidate for cultivation as transplanting is seldom successful. Correll (1950) reported that attempts at cultivation failed after a few years. The coralloid structures sometimes found on the corms suggest a need for a permanent fungal relationship that may be lost with transplanting. Another possible reason for failure in cultivation is that the plants are relatively short-lived. According to Case (1987), individual plants exist in nature for only about 5 years, with most plants living for a shorter time. A plant is vigorous for only a couple of years, after which it decreases in vigor then disappears entirely. Ashmore (1995) reported success with the artificial germination of a hybrid between *C. bulbosa* var. *americana* and *C. bulbosa* var. *occidentalis*, so commercial production of seedlings may not be too far away, and some pressure on natural populations may be removed.

Coeloglossum Hartman

Handbok i Skandinaviens Flora: 329. 1820.
Etymology: The genus name is derived from a Greek word meaning "hollow tongue," referring to the shape of the spur.

Coeloglossum (see-low-glos'-um) is a monotypic genus, circumboreal in distribution. Though first described by Hartman in 1820 based on the uniqueness of the lip, it was largely ignored by American authors such as Baldwin (1884), Gibson (1905), Ames (1924), and Correll (1950), who preferred placing the lone species in the catch-all genus *Habenaria*. Starting with Luer (1975), modern American authors began to recognize *Coeloglossum* as distinct. Authors such as Case (1987), Smith (1993), Homoya (1993), and Chapman (1997) treated *Coeloglossum* as a separate genus. Support for separate treatment came from Dressler (1981, 1993), who placed *Coeloglossum* in the subtribe Orchidinae, along with *Platanthera* and *Piperia*, based on the shape of its tuberoids. Dressler placed *Habenaria* in the subtribe Habenariinae. There are two named varieties of *C. viride*. Both occur in North America, but only one is in the lower 48 states.

 ## *Coeloglossum viride* (Linnaeus) Hartman var. *virescens* (Muhlenberg) Luer
The Native Orchids of the United States and Canada: 172. 1975.

Etymology: The epithet *viride* means "green."

Synonymy:
Basionym: *Orchis virescens* Muhlenberg in Willdenow, Species Plantarum 4: 37. 1805.
Habenaria bracteata (Muhlenberg) R. Brown, in Aiton, Hortus Kewensis, ed. 2, 5: 192. 1813.
Habenaria viridis (Linnaeus) R. Brown var. *bracteata* (Muhlenberg) A. Gray, Manual of the Botany of the Northern United States, ed. 5: 500. 1867.
Habenaria virescens (Muhlenberg) Sprengel, Systema Vegetabilium 3: 688. 1826.
Platanthera bracteata (Muhlenberg) Torrey, A Flora of the State of New York 2: 279. 1843.
Platanthera viridis (Linnaeus) Lindley var. *bracteata* (Muhlenberg) Reichenbach f., Icones Flora Germanica Recensitae 13–14: 130. Pl. 435. 1851.
Coeloglossum bracteatum (Muhlenberg) Parlatore, Flora Italiana 3: 409. 1860.
Coeloglossum viride (Linnaeus) Hartman var. *bracteatum* (Muhlenberg) Richter, Plantae Europeae 1: 278. 1890.

Common names: frog orchid, satyr orchid, long-bracted orchid, bracted green orchis.

Plate 2

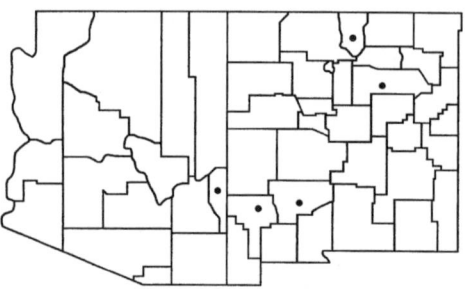

Map 2. Distribution of *Coeloglossum viride*

Description

Plant: 15 to 40 cm tall in this region, but reported as tall as 80 cm elsewhere; three to six leaves alternating along stem; inflorescence dense with 35 to 40 flowers with long floral bracts.

Roots: thick, tuberous, usually two, palmate, often with cordlike smaller roots at extreme tips of tubers and base of stem.

Leaves: 12 cm long × 5 cm wide, lower ones elliptic, becoming elliptic lanceolate higher on stem.

Floral bracts: lanceolate, to 6 cm long × 1.2 cm wide; lower bracts much longer than flowers.

Flowers: green with yellow and rose highlights on lip, 1.0 to 1.8 cm long × 0.4 to 1 cm wide; sepals and petals form hood over column.

Sepals: dark green, lateral sepals ovate lanceolate, 8 × 4 mm; dorsal sepal lanceolate, 7 × 3 mm.

Petals: pale green to whitish green, lanceolate, 4.5 × 1 mm.

Lip: pale green to yellowish green with rose to reddish shading near column, 1.0 cm long × 0.4 cm wide; oblong, spreading slightly on lower two-thirds; apex with two prominent lateral lobes connected by smaller middle lobe or tooth; thickened ridge down center.

Spur: whitish, scrotiform, 2 mm wide × 2.2 mm long.

Column: stocky, 2 mm high × 2.5 mm wide.

Capsule: elliptical, semierect to erect.

Coeloglossum viride (vir'-i-dee) var. *virescens* is the varietial name Luer (1975) used to distinguish the plants occurring in the United States and most of Canada from the European variety originally described in 1753 by Linnaeus as *Satyrium viride*. The primary differences separating *Coeloglossum viride* var. *virescens* from *C. viride* var. *viride* are the size of the plants and the size of the floral bracts. *Coeloglossum viride* var. *virescens* is the more robust of the two, growing to 80 cm tall, with floral bracts often more than three times longer than the flower. The typical form, *Coeloglossum viride* var. *viride*, seldom reaches 30 cm tall and is usually much shorter, blooming when as short as 5 cm. Its floral bracts are at most twice as long as the flower, and usually shorter. In Arizona and New Mexico, *C. viride* var. *virescens* is not nearly as robust as in other parts of the United States. It is between 15 and 40 cm tall, with most under 30 cm. However, the length of the floral bracts is still characteristic of *C. viride* var. *virescens*.

The flower of *C. viride* var. *virescens* has an interesting blend of colors and shapes. The dark green sepals and the pale green petals form a tight hood over the column. The lip, protruding prominently out from the hood, has three lobes with a central thickening, and changes from green at the apex to yellowish, with a suffusion of red near its base. On some flowers red dominates the upper half of the lip. Luer (1975) reported nectar secretions on the basal portion of the lip in recesses formed by the curved margins. The large lip and hood effectively shield the pouch-shaped spur, which is more white than pale green.

In the eastern parts of the United States, *Coeloglossum viride* var. *virescens* is often confused with *Platanthera flava* var. *herbiola* because the plants look so much alike. *Platanthera flava* var. *herbiola* does not occur in the Southwest, so that problem of mistaken identity is avoided. However, *Platanthera purpurascens*, which grows in the same areas and habitat, bears a close resemblance to *C. viride* var. *virescens* and may be confused with it. The two can be distinguished based on the structure of the leaves and lips. The leaves of *C. viride* var. *virescens* are elliptical, while those of *P. purpurascens* are linear. The lip of *C. viride* var. *virescens* is three-lobed at the apex, while the lip of *P. purpurascens* is acute rather than lobed at the apex.

Baldwin (1884) gave a detailed report on the cross-pollination of *C. viride* var. *virescens*. The opening to the saccate spur is protected by a narrow slit, which requires some probing for the insect to gain access. The insect is guided to the proper spot by the curved margins of the lip. Pollinia carried by the visiting insects come in contact with the stigma and are deposited. As the insect backs out or forages on the nectar secreted on the lip, it picks up new pollinia. The pollinia adhere to the head of the insect and after several minutes rotate to point slightly forward in order to contact the stigma on the next flower visited. The time taken for the pollinia to rotate forward helps ensure they will be deposited on a different plant than the one from which they were taken.

Distribution

Coeloglossum viride var. *viride* is circumboreal in distribution but in North America is found only in the subarctic tundra regions of Canada and Alaska. Luer (1975) showed the distribution of *C. viride* var. *virescens* as stretching from the northeastern United States and Canada across to southern Alaska and the northeastern Pacific coast of Asia. It does not occur in Europe. In the eastern United States it extends as far south as North Carolina. It is found in Washington State in the Pacific Northwest, and Case (1987) reported it is common in parts of the Great Lakes region. It also is found in the Rocky Mountain states, with disjunct populations in Utah, Arizona, and New Mexico. Herbarium records show it only from one region in Greenlee County in Arizona, but Epple (1995) published a photograph of it reportedly taken in Apache County. Its occurrence in Arizona is the southwestern limit of the species. In New Mexico, *C. viride* var. *virescens* is in Grant, San Miguel, Sierra, and Taos Counties.

Habitat

Coeloglossum viride var. *virescens* grows in a narrow elevation band between 9000 and 10,000 feet (2700 and 3048 meters). In Arizona it grows in mixed aspen and fir forest among ferns. The orchid is about the same height as the ferns and very difficult to see. In New Mexico it also grows in aspen and fir forest but in slightly more open habitat than in Arizona. The forest floor is densely populated with grasses and other small herbaceous plants that provide an effective cover for the orchids. Topography ranges from flat to gently sloping hillsides. In both Arizona and New Mexico, *C. viride* var. *virescens* grows in light to medium shade.

Blooming Season

Coeloglossum viride var. *virescens* blooms between late June and the middle of July, with peak bloom occurring near the first of July. The saccate spur fades within a few days of opening, but the rest of the flower persists for several weeks, appearing fresh even as the capsules mature. According to Reddoch and Reddoch (1997), the shoot that will be the next year's plant appears 1 to 2 cm above ground in late fall next to the old stem, and overwinters in that exposed condition. Growth resumes the following spring.

Associated orchids in flower at the same time as *C. viride* include *Cypripedium parviflorum*, *Platanthera huronensis*, *P. purpurascens*, and *Listera cordata*. Earlier bloomers in the same area include *Corallorhiza maculata*, *C. striata*, *C. wisteriana*, *Calypso bulbosa*, and *Schiedeella arizonica*. *Goodyera oblongifolia* and *G. repens* will be spiking but are not yet in bloom when the frog orchid's flowers fade.

Conservation

While *Coeloglossum viride* var. *virescens* is widely distributed and fairly common elsewhere, it is the rarest orchid in Arizona. It has a state ranking of G5/S1, which recognizes it as secure elsewhere but also means it has fewer than six occurrences in Arizona. There are only three different herbarium records of it from Arizona, the first one dating from 1937. The first literature report of it in Arizona was by D. J. Pinkava (Pinkava et al., 1975). All collections from Arizona, and the author's field observation in 1995, were from the same area in Greenlee County. Epple's report (1995) of it from Apache County suggests that *Coeloglossum viride* var. *virescens* may be more widespread in Arizona than the scientific record shows, but is merely underreported. *Coelo-*

glossum viride var. *virescens* is rare in New Mexico but occurs in several counties and in larger numbers than in Arizona. It is not on the state's list of endangered orchids.

Notes and Comments

The history and variation of common names is an interesting adjunct to the study of orchids. "Frog orchid" is one of those intriguing names that makes one wonder about its origin. Grier (1984) speculated the name is due to the resemblance of the flower to a leaping frog. The lip with its lobed apex suggests the hind legs, and the hood over the column suggests the frog's body.

One potential reason for its relative rarity in Arizona may be that *C. viride* var. *virescens* is only an occasional visitor in this extreme limit of its range. *Coeloglossum viride* var. *virescens* is known for blooming for a few years then disappearing. Reddoch and Reddoch (1997) documented blooming patterns for *C. viride* var. *virescens* in the area of Ottawa, Canada. The longest record for a plant reappearing was 4 years in a row. They believed the plants had died rather than gone temporarily dormant. Hence, the apparent scarcity of *C. viride* var. *virescens* in Arizona may be because it blooms for a year or so and dies, and it then may take multiple years for seedlings to reach blooming size or even for the area to be repopulated with seeds dispersed from other populations.

Corallorhiza Gagnebin

Acta Helvetica Physica-Mathematico 2: 57. 1755.
Etymology: *Corallorhiza* derives from Greek and means "coral root", in reference to the resemblance of the rhizome to coral.

Corallorhiza (ko-rall'-oh-rye-za) is a mycotrophic orchid genus of 11 species widely distributed in North America, with some species in Central America and one in Europe. Seven species grow in the United States and Canada; 3 of them occur in Arizona and 4 in New Mexico. The structure of the rhizome varies somewhat, but in several species it resembles coral, hence, the plants' common name.

Mycotrophic plants rely on a relationship with a mycorrhizal fungus for nutrients. All orchids are mycotrophic for at least part of their life, primarily during germination and early seedling growth. Many orchids eventually become self-supporting by photosynthesis. However, others, including most temperate zone terrestrial orchids, maintain a partial mycotrophic dependence all their life and perish if the fungus dies, even though they also use photosynthesis. Some genera, such as *Corallorhiza*, do not support significant amounts of photosynthesis and are totally or nearly totally dependent on the fungus. Without photosynthesis there is no need for chlorophyll or leaves to hold chlorophyll toward the sun. As a result, the residual leaves on *Corallorhiza* are reduced to mere sheaths on the flower stem, and the plants essentially are composed of a rhizome, stem, and flowers, appearing above ground only to bloom. Nearly totally lacking chlorophyll, the plants are mostly devoid of green pigmentation and instead exhibit many shades of browns, reds, and yellows. For a detailed discussion of mycotrophic plants and their relationship with fungi, see the books by Rasmussen (1995) and Arditti (1992).

Corallorhiza is sometimes confused with *Hexalectris*, the other mycotrophic orchid genus in Arizona and New Mexico. However, they can be distinguished easily by characteristics of the lip. The lip in *Hexalectris* has five or more lamellae down most of the central lobe. The lamellae in some species such as *H. warnockii* contrast in color to the rest of the lip. *Corallorhiza* has at most two calli and those only on the basal half of the lip, and they are the same color as the rest of the lip. Greenwood (1981) wrote that both *Corallorhiza* and *Hexalectris* may be indirect parasites, linked to a host via a mycorrhizal fungus, and he called them *mycoparasites*.

For slightly more than the last half of the twentieth century (according to Freudenstein (1996), mainly owing to a paper by Fernald (1946)), *Corallorhiza* was attributed to Chatelain based on a small monograph he published in 1760. Recent authors such as McVaugh (1985) and Freudenstein (1992, 1997) gave credit for describing the genus to Gagnebin, who published it in 1755, although Gagnebin spelled the name as *Corallorrhiza*. Freudenstein (1996) advocated that the spelling *Corallorhiza* be conserved, primarily because of its widespread use in the twentieth century.

Key to the Species of *Corallorhiza*

1. Flowers with minute to well-defined mentum, two free lamellae
    2. Lip with two small lateral lobes
        3. Sepals one-nerved.   *C. trifida*
        3a. Sepals three-nerved.   *C. maculata*
    2a. Lip without lateral lobes.   *C. wisteriana*
1a. Flowers without mentum, lamellae fused.   *C. striata*

Corallorhiza maculata (Rafinesque) Rafinesque
American Monthly Magazine and Critical Review 2: 119. 1817.

Etymology: *Maculata* is from the Latin word for "spotted."

Synonymy:
Cladorhiza maculata Rafinesque, American Monthly Magazine and Critical Review 1: 429. 1817.
Corallorhiza multiflora Nuttall, Journal Academy of Natural Sciences, Philadelphia 3: 138. 1823.
Corallorhiza mexicana Lindley, Genera and Species of Orchidaceous Plants: 534. 1840.
Neottia multiflora (Nuttall) Kuntze, Reviso Generum Plantarum 2: 674. 1891.
Corallorhiza grabhamii Cockerell, Torreya 3: 140. 1903.

Common names: spotted coralroot, large coralroot, many-flowered coralroot.

Plates 3, 4

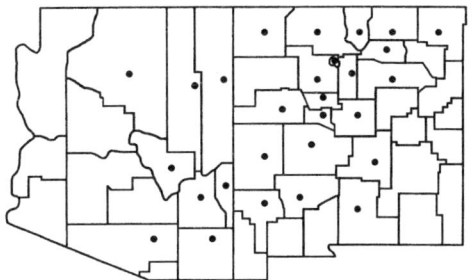

Map 3. Distribution of *Corallorhiza maculata*

Description

Plant: leafless, mycotrophic, 10 to 51 cm tall, with little or no chlorophyll, usually shades of brown, tan, pink, red, yellow, or purple.

Roots: rootless, grows from underground coralloid rhizomes.

Leaves: none at any growth stage; residual leaves reduced to sheathlike bracts totally appressed to stem; same color as stem.

Scape: brown, tan, pinkish, reddish, yellow, or purple; never green, glabrous with several sheathing tubular bracts.

Floral bracts: approximately 1 mm high × 0.9 mm wide.

Flowers: up to 50 flowers, but usually 15 to 30, each 1.8 × 1.7 cm on pedicels about 3 mm long; sepals and petals spreading.

Sepals: to 1.4 cm long × 0.4 cm wide, 3-nerved, linear to slightly elliptic, lateral sepals often slightly falcate; often lightly veined and spotted; brown, tan, pinkish, reddish, yellow, or purple; lateral sepals unite to form small mentum that extends along ovary at base of lip.

Petals: to 1.5 cm long × 0.7 cm wide, linear to slightly oblanceolate with acute apex, often lightly veined and spotted, brown, tan, pinkish, reddish, yellow, or purple.

Lip: 0.7 to 1.0 cm long × 0.4 to 0.8 wide, white, usually but not always with reddish to purplish spots; unevenly three-lobed, with small lateral lobes pointing forward; central lobe with either broadly spreading or parallel sides; central lobe to 0.8 cm wide when spreading, only about 0.4 cm wide with parallel sides; front edge of lip with slightly undulate margins; two prominent callosities at top one-half to one-third of lip.

Column: 5 mm high × 1.5 mm wide; yellow, often with a few purple to reddish spots; curved, narrower at middle than at either end; small basal lobes of column spread out and curl around to form narrow opening; terminal anther cap yellow with or without a few red or purple spots; four yellow pollinia.

Capsule: 0.6 to 2 cm long × 0.3 to 0.5 cm wide, ellipsoidal, pendent.

Corallorhiza maculata (mak-yoo-lah'-ta) arises from a coralloid rhizome as a leafless stem, with multiple bracts. The flowers, as many as 50, are the same rich brown, yellow, and red tones as the stem. The sepals and petals are often lightly spotted and faintly veined. The perianth is sometimes slightly cupped but is often free and spreading. The common name of spotted coralroot derives from the white three-lobed lip, which usually is dotted with a few to many reddish or purplish spots. However, *C. wisteriana* also has dots on the lip, and the lip of *C. maculata* is often an unspotted pure white, so it is best to rely on the three lobes of the lip and the veins in the petals for identification. The column is often sprinkled with dots, and there is a spurlike mentum formed by an extension of the sepals along the ovary.

Freudenstein (1997) recognized three varieties of *C. maculata*, and plants corresponding to all three varieties occur in Arizona and New Mexico. The three varieties can be distinguished by characteristics of the lip. *Corallorhiza maculata* var. *maculata* has parallel or nearly parallel sides on the middle lobe of the lip. Both *C. maculata* var. *occidentalis* (Lindley) Ames and *C. maculata* var. *mexicana* (Lindley) Freudenstein have broadly expanded central lobes of the lip and are differentiated by the pattern of the spots and size of the mentum. Spots on *C. maculata* var. *occidentalis* cover all portions of the lip, while those on *C. maculata* var. *mexicana* are only near the lamellae and at the edges of the central and lateral lobes. The mentum is slightly larger on *C. maculata* var. *mexicana*. Until Freudenstein's treatment of the genus, *C. mexicana* Lindley was considered a separate species, but he reduced it to a variety of *C. maculata*.

The color of *C. maculata* varies widely, even within a local area. Several varieties of *C. maculata* have been described based on color, although current botanical usage suggests they should have been designated forms instead of varieties. Correll (1950) believed that even the more striking variants were not worthy of taxonomic consideration. Freudenstein (1992, 1997) agreed with Correll, suggesting the genes for the color variations are present in the entire population, and therefore the plants are not recognizable as varieties.

The primary color forms encountered in the literature are *flavida*, *intermedia*, *punicia*, *fusca*, and *immaculata*, and all these color forms are found in Arizona and New Mexico. *Corallorhiza maculata* Rafinesque var. *flavida* (Peck) Cockerell, commonly called the yellow coralroot, applies to plants of lemon yellow with yellow flowers except for pure white unspotted lips. This color variant is the most strikingly beautiful one and is visible from many yards away if spotlighted by rays of sunlight. *Corallorhiza maculata* var. *immaculata* Peck refers to normally colored flowers with white unspotted lips. Flowers with white unspotted lips are occasional in most areas of the range and are the dominant form in some locations. *Corallorhiza maculata* var. *punicia* Bartlett has bright reddish purple stems with pure white or brightly spotted lips. The bright color of the stem and bracts extends to the backside of the sepals and petals although the front sides are much paler. *Corallorhiza maculata* var. *fusca* Bartlett is pale brown. *Corallorhiza maculata* var. *intermedia* Farwell is, as the name implies, intermediate in color between *C. maculata* var. *punicia* and *C. maculata* var. *flavida*.

Taylor and Bruns (1997) found that *C. maculata* was dependent on a fungus that was symbiotic with a host tree. They suggested the orchid was a "cheater" in the system because it provided nothing to the partnership and was therefore an "ectomycorrhizal epiparasite." This research adds to a growing body of evidence that some orchids may use their mycorrhizal fungus as a means to obtain support from other plants.

Distribution

Corallorhiza maculata is widely distributed in North America. It ranges from Canada into Mexico, and is the most common coralroot and probably the most common orchid in Arizona and New Mexico. Within Arizona it occurs in 8 counties and in New Mexico it is in 20 counties. Habitat and distribution patterns suggest it will probably be found in additional counties in both states.

Habitat

Corallorhiza maculata usually inhabits dry, open forest, at elevations between 6900 and 10,000 feet (2100 and 3000 meters). A few plants occur as low as 5250 feet (1600 meters) and as high as 10,500 feet (3200 meters) but are rare at those extremes of their elevation range. The spotted coralroot grows in the leaf litter of conifers, aspen, and oaks and on rotting logs and tree stumps in bright light to medium shade. Large colonies develop via clonal propagation by branching of the rhizome. The spotted coralroot shows some tolerance for varied growing conditions, often occupying moist environments of creek banks or river bottoms in competition with other herbaceous plants. Individuals may grow under low-growing shrubs or in open areas of the forest or on the edges of meadows in partial shade.

Blooming Season

The first spikes of *C. maculata* emerge in late April and early May, and others continue to emerge over several weeks. The buds are totally protected within the sheaths, but the colored spikes give hints of the rich hues of the flowers they will bear later. The normal blooming season stretches between late May and mid-July. In some years flowering begins in mid-May and in other years will last until late July. In most areas of Arizona and New Mexico, *C. maculata* var. *occidentalis* blooms first. While it is in flower, the spikes of *C. maculata* var. *maculata* appear, and the narrow-lip flowers open after the wide-lip plants are in fruit. The beginning of the season therefore features *C. maculata* var. *occidentalis* and the end of the season *C. maculata* var. *maculata*. During some years blooming periods for the two varieties may overlap slightly, but more often they do not. *Corallorhiza maculata* var. *mexicana* blooms in the early part of the season with *C. maculata* var. *occidentalis*.

Many other orchids grow in or near the same habitats as *C. maculata* including *C. striata*, *C. wisteriana*, *C. trifida*, all 4 *Malaxis* species, both *Goodyera* species, *Platanthera purpurascens*, *P. limosa*, *Schiedeella arizonica*, and *Cypripedium parviflorum*.

Conservation

Because of its wide distribution and adaptability to varied habitats, *C. maculata* is not threatened within Arizona and New Mexico. Significant portions of its range are protected within designated wilderness areas.

Notes and Comments

Catling (1983) reported that *C. maculata* is self-pollinating by rotation of the pollinia 270 degrees onto the stigma. Self-pollination is made possible by rapid withering of the anther cap within 2 days of the flower opening. Catling suggested that prior to the withering of the anther cap, pollination through outcrossing is possible, but specific pollinators have not been identified. The percentage of fruit set is very high, nearing 100 percent in most areas, most years.

Corallorhiza striata Lindley
Genera and Species of Orchidaceous Plants: 534. 1840.

Etymology: *Striata* is from the Latin word for "striped."

Synonymy:

Corallorhiza macraei A. Gray, Manual of Botany, ed. 2: 453. 1856.

Corallorhiza bigelovii Watson, Proceedings American Academy of Arts and Sciences 17: 275. 1877.

Neottia striata (Lindley) Kuntze, Reviso Generum Plantarum 2: 674. 1891.

Corallorhiza vreelandii Rydberg, Bulletin Torrey Botanical Club 28: 271. 1903.

Corallorhiza ochroleuca Rydberg, Bulletin Torrey Botanical Club 31: 402. 1904.

Corallorhiza striata Lindley var. *vreelandii* (Rydberg) L. O. Williams, Annals Missouri Botanical Garden. 1934.

Common name: striped coralroot.

Plates 4, 5

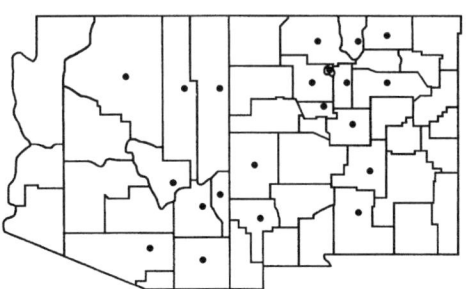

Map 4. Distribution of *Corallorhiza striata*

Description

Plant: leafless, mycotrophic, glabrous, 10 to 40 cm tall, with little or no chlorophyll, reddish purple, tan, or yellow.

Roots: none; rhizome more or less coralloid, but not always resembling coral.

Leaves: none at any growth stage.

Scape: few to 30 flowers; several tubular bracts along stem; reddish purple, tan, or yellow.

Floral bracts: triangular, about 3 mm at base × 4 mm long, same color as scape.

Flowers: 1.0 to 1.5 cm across × 0.9 to 1.2 cm high, sepals and petals connivent to spreading; reddish purple, tan, or yellow, usually with stripes on sepals, petals, and lip; stripes sometimes not present in yellow forms.

Sepals: dorsal sepal elongated elliptic to lanceolate, 0.4 cm wide × 0.95 cm long; lateral sepals lanceolate with rounded apex, slightly falcate, slightly concave; 0.35 cm wide × 1.1 cm long; reddish purple, tan, or yellow, usually with three well-defined stripes in dark red, purple, or yellow.

Petals: elliptic oblanceolate, 0.4 cm wide × 1.0 cm long, reddish purple, tan, or yellow with five darker stripes, two of which are on opposite edges.

Lip: elliptic ovate, entire, without mentum, edges recurved upward, creating shallow cup shape; 0.5 cm wide × 0.8 cm long; reddish purple, tan, or yellow with five stripes in dark red, purple, or yellow, wider than stripes on sepals and petals; two lamellae fused for part of length on basal part of lip, often appearing as single bilobed callus.

Column: curved, about 0.5 cm long × 0.25 cm wide, yellow, narrower at middle; with terminal anther cap and four yellow pollinia.

Capsule: ellipsoidal, pendent, 1 to 1.6 cm long × 0.5 to 1 cm wide, with faint striping.

Corallorhiza striata (stry-ay'-ta) is very easy to distinguish from the other coralroots. It is profusely striped on all parts of the perianth, it does not have a mentum, and the lip is entire with smooth margins. As many as 30 flowers bedeck the leafless stem. The striped dorsal sepal and petals are held closely together behind the column. The striped lateral sepals are somewhat spreading, although all sepals and petals are slightly concave. The lip is boat-shaped, with five bold stripes, including one along each turned-up margin. Freudenstein (1997) recognized three varieties of *C. striata* based on sizes of floral parts. *Corallorhiza striata* var. *striata* has the largest flowers, with lips over 7 mm long. *Corallorhiza striata* var. *vreelandii* (Rydberg) Williams and *Corallorhiza striata* var. *involuta* (Greenman) Freudenstein have smaller flowers, with lips under 7 mm long. The latter two varieties are differentiated by the width of the lip. The former has lips wider than 2 mm, and the latter has lips narrower than 2 mm. *Corallorhiza striata* var. *involuta* usually had been considered a separate species, *C. involuta* Greenman, but Freudenstein included it within his expanded concept of *C. striata*. All *C. striata* in both Arizona and New Mexico fall within the size limits of *C. striata* var. *vreelandii*. Rydberg originally named *C. vreelandii* in honor of a field companion.

At least two color varieties of *C. striata* have been named. *Corallorhiza striata* forma *fulva* Fernald has sheaths and perianth of yellow or orange-brown but still has pronounced stripes. *Corallorhiza striata* forma *fulva* is the most common form of *C. striata* in New Mexico according to Todsen and Todsen (1971), although in the author's experience, typical *C. striata* var. *vreelandii* is by far more common. Yellow plants and flowers without stripes have been called either *C. striata* var. *flavida* Todsen and Todsen, or *C. ochroleuca* Rydberg.

Freudenstein (1997) reported that *C. striata* is most likely pollinated by parasitic wasps, *Coccygonimus pedalis*. The wasps were observed visiting flowers and transporting pollinia. Self-pollination also occurs, but at a much lower rate than the outcrossing accomplished by the wasps.

Distribution

Corallorhiza striata is widely distributed, ranging from California into Canada and across the northern states, with a disjunct population in Mexico. Within Arizona it occurs in the 8 counties of Apache, Cochise, Coconino, Gila, Graham, Greenlee, Navajo, and Pima. In New Mexico it occurs in the 13 counties of Bernalillo, Catron, Colfax, Grant, Lincoln, Los Alamos, Otero, Rio Arriba, San Miguel, Sandoval, Santa Fe, Taos, and Torrance. It grows in the Guadalupe Mountains in Texas just south of Eddy County, New Mexico. Since the Guadalupe Mountains extend into Eddy County, New Mexico, *C. striata* may also grow there.

Habitat

The striped coralroot grows in evergreen and mixed deciduous forests at elevations between 6900 and 9500 feet (2100 and 2900 meters). It grows predominantly in dry, open forest on flat to steep terrain in bright to moderate shade. Sometimes it grows beneath a light overstory of shrubs, and less often it grows in moist environments near streams.

Blooming Season

Spikes of *C. striata* first appear in late April to early May, and at this stage, still tightly in sheath, it is impossible to determine which species of coralroot it is. Later, as the buds begin to differentiate, the stripes make identification easy. The primary blooming season is between the middle of May and the end of June. If spring is early, flowering may start by the first week of May, and in years with late springs or cool summers it may last until the middle of July.

Individuals do not bloom every year, often skipping one or more years before again appearing above ground. Reddoch and Reddoch (1997) reported on a study lasting 29 years on a single colony of *C. striata*, during which the number of blooming plants varied from 0 to 155. There were no blooming plants in 4 nonconsecutive years during the study.

Corallorhiza striata often flowers in the immediate vicinity of *C. maculata* and *C. wisteriana*. Other orchids blooming nearby include *Malaxis soulei, Platanthera purpurascens, Goodyera oblongifolia, G. repens*, and *Schiedeella arizonica*.

Conservation

Corallorhiza striata is widespread in Arizona and New Mexico. Portions of its habitat are protected within designated wilderness areas and are safe from development.

Notes and Comments

While the sizes of all *C. striata* in Arizona and New Mexico fall within the definition of *C. striata* var. *vreelandii*, it is the author's experience that plants intermediate in size between the two supposed varieties exist in other places in its range. Therefore, *C. striata* var. *vreelandii* is treated as a synonym of *C. striata*.

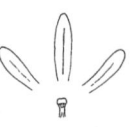

Corallorhiza trifida Chatelain
Specimen Inaugurale de Corallorhiza: 8. 1760.

Etymology: *Trifida* derives from the Latin word for "split into three," in reference to the three-lobed lip.

Synonymy:
Corallorhiza innata R. Brown, in Aiton, Hortus Kewensis, ed 2, 5: 209. 1813.
Corallorhiza verna Nuttall, Journal Academy of Natural Sciences, Philadelphia 3: 136. 1823.
Corallorhiza corallorhiza Karsten, Deutsche Flora: 488. 1881.
Corallorhiza innata R. Brown var. *virescens* Farr, Transactions Botanical Society of Pennsylvania 2: 425. 1904.
Corallorhiza corallorhiza Karsten ssp. *coloradensis* Cockerell, Torreya 16: 231. 1916.
Corallorhiza wyomingensis Hellmayr & Hellmayr, Rhodora 33: 133. 1931.
Corallorhiza trifida Chatelain var. *verna* (Nuttall) Fernald, Rhodora 48: 196. 1946.

Common names: early coralroot, pale coralroot, northern coralroot.

Plate 5

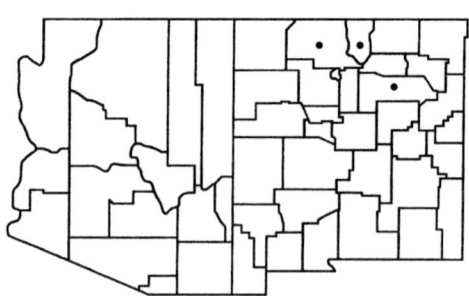

Map 5. Distribution of *Corallorhiza trifida*

Description

Plant: leafless, mycotrophic, glabrous, 10 to 25 cm tall, greenish or yellow with 10 to 20 flowers.

Roots: none, rhizome more or less coralloid.
Leaves: none at any growth stage.
Floral bracts: barely noticeable, about 1 mm long.
Flowers: green with white lip, 1 cm across.
Sepals: green, lateral sepals linear lanceolate, one-nerved, falcate, 4 × 1 mm; dorsal sepal oblanceolate, 3 × 1 mm.
Petals: green, oblanceolate, 3 × 1 mm.
Lip: white, without dots here but often spotted elsewhere, about 5 × 3 mm, three-lobed, but lateral lobes minute, central lobe with margins nearly parallel to slightly expanded on lower half, rounded at apex; two raised calli in upper third of lip and margin of lip is ruffled.
Column: green, curved, about 4 mm long, pollinia yellow.
Capsule: ellipsoidal, pendent.

Corallorhiza trifida (trif'-id-ah) is the smallest of the coralroot species in Arizona and New Mexico. Mature plants are typically under 20 cm tall and bear about 10 flowers. Color is a great aid to identification. The leafless raceme, sepals, and petals are a uniform chartreuse green to very yellowish green, and the lip is unspotted white. However, neither color nor size should be the primary method of identification because the height of *C. trifida* overlaps that of *C. maculata* and *C. wisteriana*, and yellow forms of both the latter two species are occasionally encountered. Rather the shape of the lip and lateral sepals separate *C. trifida* from the others. The pure white lip is three-lobed as in *C. maculata*, but the lateral lobes of the lip in *C. trifida* are not as long, appearing more as a tooth or notch. The lip of *C. wisteriana* is entire, without lobes. The linear-lanceolate lateral sepals extend below the lip in a slight curve and are proportionally longer than in *C. maculata* or *C. wisteriana*. The petals are oblanceolate and remain partially closed about the column, a characteristic common in *C. wisteriana*, but usually not seen in *C. maculata*. The greenish coloration of *C. trifida* means that unlike other coralroots, chlorophyll is present in relatively large amounts.

The typical color form of *C. trifida* found in the northeastern United States is often slightly tinged with brown on the sepals and petals and may have purple spots on the lip. The light green flowers with pure white lips such as those that grow in New Mexico were originally segregated as *C. verna* Nuttall, but were subsequently reduced to a variety as *Corallorhiza trifida* Chatelain var. *verna* (Nuttall) Fernald. In his monograph of the genus, Freudenstein (1997) did not recognize var. *verna*, stating that intermediate forms exist, and in parts of the range both color forms grow side by side. The inclusive nomenclature recommended by Freudenstein is followed here.

Recent research is providing more information on how mycotrophic plants such as *C. trifida* obtain nutrients. Zelmer and Currah (1995) stated that "*Corallorhiza trifida* may be tapping into a mutualistic symbiosis between the ectomycorrhizal yellow fungus and its tree hosts, and perhaps acting as a sink for the tree's photosynthates via the fungus." In effect, the orchid is an indirect parasite on the tree (lodgepole pine, *Pinus contorta*, in the study) via the fungus.

Catling (1983) reported that *C. trifida* in North America exhibits a high percentage of self-pollination. This is consistent with Davies, Davies, and Huxley's (1988) observation that European plants are self-pollinated when the pollinia fall onto the stigmatic surface.

Distribution

The northern coralroot is well known because it is circumpolar in distribution and is quite common in parts of its range, occurring in large colonies. *Corallorhiza trifida* is the only member of the genus occurring in Europe and Asia. Although the most common coralroot in Canada and found fairly frequently in the eastern and northern parts of the United States, *C. trifida* is spotty in its distribution in the western United States. While widespread in Colorado and Wyoming, only about a dozen sites are known from Utah. It occurs in only one location in Nevada, is rare in the Blue and Wallowa Mountains of Oregon, and is one of the rarest orchids in California. It is not known from Arizona but occurs infrequently in New Mexico in Rio Arriba, San Miguel, and Taos Counties, where it is at its southern limit.

Habitat

Corallorhiza trifida grows in a variety of habitats at elevations between 9000 and 10,500 feet (2740 and 3200 meters). At the lower end of its elevation range it grows in damp areas in canyon bottoms, usually in proximity to a stream or river. It is also secluded in damp areas of pine and fir forest among mosses and often is partially hidden under the branches of low-growing shrubs. At higher elevations it occupies more mesic areas of the forest; at the upper limits of its elevation range it grows in dry pine and fir duff in the open forest. This habitat range is similar to that observed by Reddoch and Reddoch (1997). They reported *C. trifida* var. *trifida* from swamps and wet ground but said that *C. trifida* var. *verna* grows in mixed and coniferous forest of mesic habitat.

Blooming Season

Corallorhiza trifida begins flowering during the middle of June and continues into late July. Although known as the early coralroot because it is the first of its genus to bloom in other parts of the United States, *C. trifida* is the last coralroot to open in New Mexico. This change in blooming sequence is probably due to the relatively high elevations where it grows. The other three members of the genus in these two states —*C. maculata, C. striata,* and *C. wisteriana*—grow at lower elevations, where they flower earlier. When at the same elevation, *C. maculata* and *C. striata* come into bloom at approximately the same time as *C. trifida.*

Other orchids blooming nearby may include *Cypripedium parviflorum, Calypso bulbosa, Coeloglossum viride, Corallorhiza maculata, C. striata, Platanthera purpurascens,* and *Listera cordata. Goodyera oblongifolia* grows in the same area but will not bloom until after *C. trifida* fades.

Conservation

Corallorhiza trifida is relatively rare in New Mexico and historically was known from three collections made between 1908 and 1956. A fourth location was discovered in 1999. All locations are on National Forest lands, and some are within the protection of wilderness areas, where they are presumably safe from developmental threats. *Corallorhiza trifida* is not currently on the list of endangered orchids in New Mexico, but it clearly belongs there owing to the relatively few occurrences in the state.

Corallorhiza wisteriana S. W. Conrad

Journal Academy of Natural Sciences, Philadelphia 6: 145. 1829.

Etymology: This orchid is named after American botanist Charles J. Wister, who collected the type specimen.

Synonymy:
Corallorhiza unguiculata, Rafinesque, Herbarium Rafinesquianum: 75. 1833.
Corallorhiza odontorhiza, Chapman, Flora of the Southern United States: 454. 1860. Not Nuttall.

Common names: spring coralroot, Wister's coralroot.

Plates 6, 7

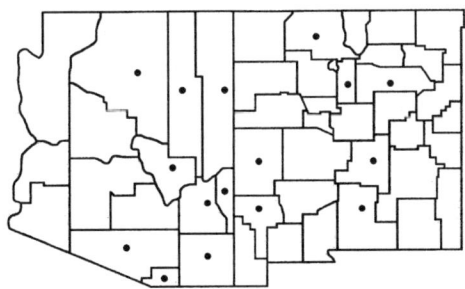

Map 6. Distribution of *Corallorhiza wisteriana*

Description

Plant: leafless, mycotrophic, glabrous, 6.5 to 38 cm tall, dark brown to tan, purplish, yellow, or greenish; frequent hints of green in plants suggest significantly more chlorophyll than in *C. maculata* or *C. striata*.
Roots: none, from branched coralloid rhizome.
Leaves: none, reduced to bracts on raceme.
Scape: dark brown to tan, purplish, yellow, or greenish; with two to four tubular bracts, uppermost bract often free of stem for 2 cm; few to 32 flowers on upper 4 to 13 cm.
Floral bracts: minute to 2 mm long, triangular to lanceolate, same color as stem.

Flowers: 1.6 cm wide × 1.8 cm high on slender pedicels up to 3 mm long; sepals and petals connivent to spreading.

Sepals: tan, purplish, yellow, lighter with some green toward base; lateral sepals lanceolate and slightly falcate, 1.0 × 2.5 cm; bases of lateral sepals combine to form mentum under column; dorsal sepal linear acute to ovate lanceolate, 1.0 × 0.2 cm.

Petals: tan, purplish, yellow, lighter with some green toward base; oblanceolate, 1.0 × 0.2 cm.

Lip: elliptic to ovate, 0.4 × 1.0 cm to 0.7 × 0.9 cm, white spotted with fine reddish to purplish dots to pure white; entire, without lateral lobes, with fine fringes along curved-up margins, narrows down to small claw at column; turns sharply downward on outer half; two basal lamellae 2 to 4 mm long.

Column: slightly curved, 6 mm long × 2 mm at top and bottom, narrower at middle; column may have reddish highlights along edges; winged at base, spreading out and curving around to form tiny tubelike opening; anther cap covers four disc-shaped dark maroon to purplish pollinia; stigma rectangular to nearly circular.

Capsule: ellipsoidal, 1.2 × 0.7 cm; pendent, often greenish, but also tan to purplish.

Corallorhiza wisteriana (wis-ter′-ee-a-na) is a smaller, daintier plant than either *C. maculata* or *C. striata* but usually slightly larger than *C. trifida*. The sepals and petals of *C. wisteriana* are most often tightly cupped over the column. Flowers with an open, spreading perianth are frequent, however, though both types never appear on the same plant. The lip is usually covered with fine to coarse reddish to purplish dots. The shape of the lip varies from narrowly elliptic to very ovate. The sides of the lip turn up so it is somewhat canoe-shaped, and the margins of the lip vary from finely to coarsely jagged. Like *C. maculata*, *C. wisteriana* may have pure white lips, and plants with pure white lips grow intermixed with those with normally spotted lips. The white-lipped plants can be the same shades of brown, tan, and purple as the spotted ones but also occur in shades of yellow and gold. A pure yellow form, with a white unspotted lip, analogous to the yellow form of *C. macu-*

lata, occurs occasionally among normally colored populations. It is easy to distinguish *C. wisteriana* from *C. maculata* even though both usually have spotted lips. *Corallorhiza maculata* has small lateral lobes on the margins of the lip, and *C. wisteriana* is without lobes.

Some plants of *C. wisteriana* have traces of green in the stem, and on many the ovaries turn greenish as they mature. This green coloring suggests *C. wisteriana* has small amounts of chlorophyll, although its role in food production is probably minor compared to the nutrients derived from the mycorrhizal relationship with its fungus.

Freudenstein (1997) did not identify pollinators for *C. wisteriana* but suggested the flower morphology was consistent with outcrossing. A small percentage of the plants he studied (2.8 percent) appeared to be self-pollinating. Clearly some of the plants in Arizona and New Mexico are self-pollinating and form fruit without the flowers fully opening.

Distribution

In the eastern part of the United States, *C. wisteriana* grows from Pennsylvania south to Florida and then west to Texas and Oklahoma. For many years, at least until the study of Ames (1924), it was thought to be restricted to that part of the country. Misidentification of herbarium specimens as *C. maculata* is believed to be the main cause of the error in the reported distribution. Now it is understood to also occur in the West from Montana and Wyoming south to Utah, Arizona, and New Mexico and into Mexico. Within Arizona it occurs in the 9 counties of Apache, Cochise, Coconino, Gila, Graham, Greenlee, Navajo, Pima, and Santa Cruz. Along with the populations in Utah, the *C. wisteriana* in Arizona are at its western range limit. Within New Mexico it grows in the 7 counties of Catron, Grant, Lincoln, Otero, Rio Arriba, San Miguel, and Santa Fe, and probably Los Alamos County. It grows in the Guadalupe Mountains in Texas, just south of Eddy County in New Mexico, so it may eventually be found in the Guadalupe Mountains in Eddy County.

Habitat

Corallorhiza wisteriana grows at elevations between 5500 and 9800 feet (1675 and 3000 meters). At the lower end of that elevation range, it is found in mixed stands of juniper and oak. Slightly higher up it occurs in mixed and pure stands of oak and pine. At the upper limits of its range, it grows in pine and fir forests. In all those diverse forests, it grows on flat forest floors and on steep hillsides, usually in deep duff but also among rocks. Typically it is in moderate to light shade but sometimes is in bright light part of the day. It occurs in the dry forest areas that are home to *C. maculata* and *C. striata*, but more frequently than the other two, is found in slightly more moist spots along streams and the bottoms of canyons near stream beds. In these damper spots it grows in the open or among and under ferns. It appears as scattered individuals or in loose to dense colonies. One group of 10 identical inflorescences, suggesting they were from the same rhizome, bloomed in a roughly square area measuring about 4 cm on each side.

Blooming Season

Corallorhiza wisteriana flowers from the middle of April to about the third week in June, and not only is the first coralroot but also is one of the first orchids to open in Arizona and New Mexico. This tendency toward early bloom has a price, since late-spring snows sometimes nip plants emerging early in the season. The flower buds are not hardy, and plants that are in bud when the snow falls are killed back. In a reversal of what would appear to be normal logic, the lower-elevation plants succumb to the late snows at a greater rate than the higher-elevation ones. The lower plants normally bloom a week or two earlier than the higher plants and are more likely to have buds fully emerged from the protecting sheath. Higher-elevation plants still in sheath have a measure of protection against the cold and progress to bloom normally.

Blooming is impacted not only by the timing but also by the amount of precipitation. During the drought year of 1996, fewer than the normal number of plants appeared at the lower elevations, and many that did appear withered early or did not fully develop. Higher in the mountains where more moisture accumulated, most plants developed to flowering.

Corallorhiza wisteriana grows near *C. maculata* and *C. striata*, and in some locations it is possible to see all three in bloom from the same spot. *Schiedeella arizonica* also blooms at the same time. Later in the season, all 4 *Malaxis* species bloom nearby, as do *Platanthera limosa*, *P. purpurascens*, and *Goodyera oblongifolia*.

Conservation

Corallorhiza wisteriana is widespread and grows in habitats common in these states. Portions of its range are protected within wilderness areas, so it is considered safe from threats at this time.

Cypripedium Linnaeus

Species Plantarum 2: 951. 1753.
Etymology: *Cypripedium* is from the Greek words for "Aphrodite," the goddess of love, and "foot," and would closely translate as "Aphrodite's foot."

Cypripedium (sip-ri-pe'-dee-um) is one of four genera of slipper orchids in the subfamily Cypripedioideae. *Cypripedium* includes north temperate and Asian deciduous species. The other three generally recognized genera in the subfamily are *Selenipedium*, a small Central and South American genus of plants 4 to 5 meters tall and with trilocular ovaries; *Phragmipedium*, a Central and South American genus with conduplicate leaves; and *Paphiopedilum*, an Old World tropical genus with conduplicate leaves. *Phragmipedium* and *Paphiopedilum* are common in the horticultural trade as home and greenhouse plants.

Cypripedium is deciduous from perennial rhizomes. The plicate leaves are sometimes basal, but more often alternate or are opposite higher on the stem. Members of this genus are easily recognized owing to the distinctive shape of their flowers. The most prominent feature is the lip, which is modified into a pouch. The pouch resembles a slipper in profile and is the origin of the common names "lady's slippers" and "moccasin-flower." Most orchids have a single fertile stamen, but *Cypripedium* has two, one on each side of the column. The column has a terminal staminode, which is a third sterile anther, often in a contrasting color to the lip and petals. Another modification in *Cypripedium* is the reduction of the two lateral sepals into a single structure called the *synsepal*. The synsepal lies behind the pouch below the dorsal sepal. In some species a close inspection of the synsepal reveals a notch at the tip, which is a remnant of its derivation from two sepals.

The pouch, in addition to being the most colorful part of the lady's slipper, is an insect trap central to the pollination process. The potential pollinator first lands on and then enters the pouch, drawn by either the aroma or the lure of nectar. Unable to fly out because of the size of the opening, it eventually finds fine hairs creating a path past the stigma and one of the anthers. In the struggle to escape, the insect first deposits any pollen it may have brought with it, and then comes in contact with a new pollen mass and carries it away. The pollen is deposited in the next flower visited, effecting cross-pollination.

Cypripedium is circumboreal, and the number of described species has been increasing due to new discoveries in China. Of the 45 species recognized by Cribb (1997), 30 are from China. The country with the next greatest number of species is the United States, with 11. Only one, *Cypripedium parviflorum* var. *pubescens*, grows in Arizona and New Mexico.

Cypripedium parviflorum Salisbury var. *pubescens* (Willdenow) Knight
Rhodora 8: 93. 1906.

Etymology: *Parviflorum* means "small flowered" and may have been in reference to the larger flowers on *C. calceolus*.

Synonymy:
Basionym: *Cypripedium pubescens* Willdenow, Hortus Berolinensis 1: Pl. 13. 1804.
Cypripedium flavescens A. P. de Candolle, Les Liliacees I: Pl. 20. 1802.
Cypripedium hirsutum Miller, Memoirs Torrey Botanical Club 5: 121. 1894.
Cypripedium veganum Cockerell & Barker, Proceedings Biological Society of Washington 4: 178. 1901.
Cypripedium parviflorum var. *planipetalum* Fernald, Rhodora 28: 168. 1926.
Cypripedium calceolus Linnaeus var. *pubescens* (Willdenow) Correll, Botanical Museum Leaflets 7: 14. 1938.
Cypripedium calceolus Linnaeus var. *planipetalum* (Fernald) Victorin & J. Rousseau, Contrbutions de l'institu Botanique de l'université de Montreal 36: 68. 1940.

Common name: yellow lady's slipper, large yellow lady's slipper, whippoorwill-shoe.

Plate 8

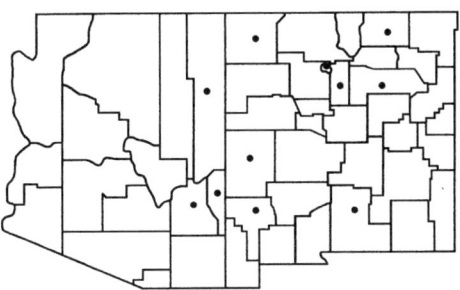

Map 7. Distribution of *Cypripedium parviflorum* var. *pubescens*

Description

Plant: 16 to 60 cm tall, reported to 80 cm elsewhere; stem, bracts, and leaves pubescent; flower stem rises 10 cm above leaves; one flower, rarely two.

Roots: few to many slender roots to 4 cm long on jointed rhizome.

Leaves: ovate lanceolate, cauline, plicate, four to six on blooming plants, from 9 × 4 cm to 14 × 5 cm; covered with fine hairs on underside; a few hairs on topside.

Floral bracts: ovate lanceolate bract at base of ovary, 7 × 2 cm.

Flowers: bright yellow pouch with greenish to reddish sepals and petals; up to 10 cm high × 10 cm wide.

Sepals: yellowish green with reddish stripes that turn to dots near pouch, fine hairs on back and edges; dorsal sepal ovate lanceolate, slightly concave, 4 × 2.2 cm; synsepal elliptic, slightly concave; slight notch at tip, 3.2 × 1.4 cm.

Petals: linear, acute, yellowish green with reddish stripes that turn to dots at pouch; 5.5 cm long × 0.7 cm wide; twisted; fine hairs on back along well-defined central ridge; few hairs on inner one-third toward pouch.

Lip: bright yellow obovoid (pouch- or slipper-shaped), 3.2 cm wide × 4.0 cm high; opening 1.2 × 2.0 cm, with incurved margin; red dotted stripes on veins and faint reddish dots on inside and back of pouch.

Column: light green with red dots at base, 1.5 cm high; two fertile anthers, one to either side; staminode yellow with red dots, arrowhead-shaped, with V form (folded); pollinia yellow sticky masses.

Capsule: ellipsoidal, pubescent, 2.2 to 3 cm long × 0.6 to 1.3 cm in diameter.

Cypripedium parviflorum var. *pubescens* (par-vi-flor'-um variety poo-bes'-enz) boasts the largest flower of any of the wild orchids in Arizona and New Mexico. The flower measures up to 10 cm from tip to tip across the petals and has nearly as long a span from tip to tip across the sepals. Most plants in this region flower when between 20 and 40 cm tall, but some get as tall as 60 cm, and a few flower when the plant is as small as 16 cm. Each year new shoots sprout from a rough rhizome, with the new bud for the next year's growth forming before the plant goes dormant. The old rhizome persists for several years. Most plants develop a single new lead every year, but sometimes the rhizome branches and a cluster of identical plants develops. Flowering-size plants have four to six ovate-lanceolate plicate leaves. The leaves, leaf bracts, floral bract, and ovaries are covered with fine hairs. Most plants bear only a single flower, though perhaps one out of 100 will have two flowers.

The yellow lady's slipper is one of a few orchids native to the Southwest that can be seen nearly 100 yards (90 meters) away if in bloom. The yellow of the pouch stands out jauntily against the greens and browns of the forest. The bright yellow pouch has an incurved margin to the opening, and the inside is marked with red dotted stripes on the veins. The yellow staminode lies just above the opening to the pouch. Pollinia on the fertile anthers are visible at either side of the staminode. Though the lady's slippers give off a pleasant sweet aroma, most authorities report that they are devoid of nectar and merely trick the pollinators with the promise of a reward. The sepals and petals are yellowish to greenish but appear tan to rich brown because of reddish stripes that are close together at the apex before becoming more widely spaced, then separating into dots near the column. Both the dorsal sepal and the synsepal are slightly concave, and the petals twist clockwise when viewed from the apex toward the base. The amount of twisting varies: on some plants the petals may be totally untwisted with wavy margins while others will have up to five complete turns.

As might be expected in large populations of flowering plants such as occur in areas of New Mexico, significant variation in color and even some structural freaks are manifest. The reddish dots that give the brownish color to the sepals and petals are totally lacking in some plants, and the sepals and petals are pure pale yellow or greenish yellow. Intermediate forms exist, so the spectrum of color in the sepals and petals varies almost continuously from dark, rich reddish brown to the pure yellow or greenish yellow. Several structural variations are common. The most often encountered is the development of two sepals instead of the usual synsepal. Other plants lack one or both petals. The color variation and the structural deformities are probably genetic characteristics carried within the population as a whole, rather than distinct varieties or forms. A two-flowered plant had a normal synsepal on the lower flower and two lateral sepals on the upper flower.

Cypripedium parviflorum has experienced a variety of treatments since Salisbury described it in 1791, especially by American authors. During the last part of the nineteenth century, Baldwin (1884) used Willdenow's *C. pubescens*. In the early part of the twentieth century, Niles (1904) and Gibson (1905) called the large yellow lady's slipper *C. hirsutum*, which is now considered a synonym for *C. parviflorum* var. *pubescens*. About the same time, the name *C. veganum* Cockerell and Barker came into use specifically for the large yellow lady's slippers in New Mexico. The species was thought to be separate from the eastern plants, based on perceived differences in color and in the shape of the stigma, but is now treated as the same. A few years later many authors such as Ames (1924) and Bingham (1939) followed Salisbury's approach, calling it *C. parviflorum* var. *pubescens*. Morris and Eames (1929) also used the name *C. parviflorum* var. *pubescens*, stating it was distinct from the European *C. calceolus*. The treatment that was to dominate most of the rest of the century was set when Correll (1950) departed from Salisbury's treatment, claiming all the yellow lady's slippers were conspecific with the European *C. calceolus*. Correll reduced the large yellow lady's slipper to varietal status (*C. calceolus* var. *pubescens*). Case (1964), Luer (1975), Summers (1981), and Smith (1993) followed Correll. Atwood (1984) departed from Correll by recognizing that the yellow lady's slippers in North America differed from the European ones at the specific level. Atwood based his decision on differences in the staminode, which in *C. calceolus* is white and obovate and in *C. parviflorum* is yellow and V-shaped. Atwood relied on Willdenow's designation, calling the large flowered yellow lady's slippers *Cypripedium pubescens*. Sheviak (1993, 1994, 1995) over a period of 3 years treated what he called the *Cypripedium parviflorum* complex. He recognized a single species with three varieties: *C. parviflorum* var. *parviflorum*, *C. parviflorum* var. *makasin*, and *C. parviflorum* var. *pubescens*. Sheviak's return to Knight's nomenclature was followed by Cribb (1997) and Keenan (1998) and is adopted here.

Part of the apparent confusion in the name is due to the extreme variation of *C. parviflorum* var. *pubescens* across its range. The size of the plant, size of the lip, size and shape of the petals, and twisting of the petals vary from site to site and within colonies. Sheviak (1995) believed the variation depends on soil type and exposure to sun. Plant structure especially depends on environmental conditions. Plants in the open have ascending narrow leaves while those in shadier conditions have leaves that are spreading and broader.

Distribution

This is perhaps the most widely distributed and most common *Cypripedium* species in North America. Cribb (1997) showed its distribution (although the key on Cribb's distribution map is mislabeled) as extending from Newfoundland to Alaska and south to Oregon in the West. In the East along the Atlantic Coast, *C. parviflorum* var. *pubescens* is in every state except Florida and extends across to Louisiana and eastern Texas. Large populations in the Rocky Mountain states are isolated by the Great Plains from those in the Great Lakes region. *Cypripedium parviflorum* var. *pubescens* occurs by the thousands in New Mexico but is rare in Arizona at the extreme southwestern limit of its range. In New Mexico it grows in Catron, Colfax, Grant, Los Alamos, Otero, San Miguel, San Juan, and Santa Fe Counties. In Arizona it is only in the counties of Apache, Graham, and Greenlee.

Habitat

Cypripedium parviflorum var. *pubescens* grows in moderate shade to nearly full sun in fir, pine, and aspen forest between 6000 and 9500 feet

(1830 and 2900 meters). The existing colonies in Arizona are near water sources such as on or near the bank of a small stream or on the edge of a seepage area. Notes on herbarium specimens indicate it also grows in mountain meadows and on timbered slopes. In New Mexico, where it is far more common, it most often grows just above the bench on the sides of drainages, usually 50 to 100 yards (45 to 90 meters) from any water. The slope varies from nearly flat up to 60 degrees. The slopes face east to northeast and are covered with lush growth under 30 cm tall, such as blueberries (*Vaccinium oreophilum*), shooting stars (*Dodecatheon pulchellum*), and several species of daisies. Another companion plant is the beautiful orange-red wood lily (*Lilium philadelphicum*), which blooms at the same time as the lady's slippers.

A second habitat in New Mexico is a significant contrast to the more common mesic forest slopes. At the low end of its elevation range, *C. parviflorum* var. *pubescens* grows in dripping seeps on steep to moderately sloped canyon walls. The soil is saturated, and the orchids are nearly hidden by a thick mat of grasses and sedges that obscure all but the flowers. The seeps are surrounded with pines and firs, but the lady's slippers are in full sun much of the day.

Blooming Season

New growth of *C. parviflorum* var. *pubescens* appears above ground in middle to late May. Flowering starts as early as the last week in May and is usually over by the first week of July. In Arizona most plants bloom in early June, but in New Mexico, where the majority of plants are at higher elevations, blooming peaks in late June and lasts to early July. The rate of fruit set in Arizona is fairly low, with only 20 to 25 percent of the flowers forming a capsule. In New Mexico, perhaps because the greater number of flowers attract more pollinators, fruits set on 50 to 75 percent of plants. Seed capsules mature by late September and early October.

Several other orchids bloom close to *C. parviflorum* var. *pubescens*. *Coeloglossum viride*, the narrow lip form of *Corallorhiza maculata*, and

Platanthera purpurascens bloom within yards and sometimes within inches of the lady's slippers. *Corallorhiza wisteriana* will already be in capsule when the lady's slippers bloom. *Goodyera repens* and *G. oblongifolia* will be in spike nearby but will not flower for several more weeks. In wet places, often within view of the *Cypripedium* plants, *Platanthhera huronensis* will be blooming.

Conservation

Within Arizona *C. parviflorum* var. *pubescens* is an extremely rare plant. Most herbarium collections were made prior to 1930, and locational data accompanying the specimens are often insufficient to relocate plants. Two colonies located by the U.S. Forest Service beginning in the 1980s have been extirpated, probably by people digging to transplant them to gardens. There are three extant colonies in Apache County. One is on private land and consists of about 15 plants in two groups. Five to 7 plants flower there each year. A second colony is on U.S. Forest Service land and in 1996 consisted of about 20 plants in three groups. In 1997 the emerging plants were browsed to the ground in early spring and did not show any additional growth that year. In 1998, ten plants in a single group came up, and only one bloomed. The other two groups did not show any sign of growth. This site is near a populated area, and a threat comes from other quarters in addition to herbivores. There is evidence of digging at the site, suggesting some plants were deliberately removed, although all three groups appeared in 1999. The third location in Arizona is on U.S. Forest Service land and consists of five to seven seedling-sized plants. They have been observed since 1995 but had not reached blooming size by 1999. A photograph from 1965 established *C. parviflorum* var. *pubescens* on the White Mountain Apache Reservation in Apache County, but that colony has not been verified as still extant. *Cypripedium parviflorum* var. *pubescens* is ranked G4/S1 within Arizona. The global ranking of G4 should more correctly be G5 because of the great numbers of plants and extensive range outside the state. The S1 ranking correctly identifies it as endangered within Arizona.

The outlook for *C. parviflorum* var. *pubescens* in New Mexico is far brighter than it is in Arizona. There the large yellow lady's slipper is widespread and locally common. A conservative estimate puts the number of plants in one area at well over 5000. In a few places it grows near the edges of public campgrounds. Even with its greater presence in New Mexico, the yellow lady's slipper is still threatened by habitat destruction and collection. Sivinski and Lightfoot (1995) believe it has probably been extirpated from Otero County, where it was last seen in 1914. The yellow lady's slipper is on List 1 in New Mexico, which provides needed legal protection against unauthorized collecting.

Notes and Comments

The beauty of *C. parviflorum* var. *pubescens* has led many gardeners to attempt to cultivate it, and colonies readily accessible to the public often suffer losses by transplanting. Fortunately, there is hard evidence that digging of wild plants for gardens is no longer necessary. The secrets of germinating most North American terrestrial orchids are still elusive, but that is not the case with *C. parviflorum* var. *pubescens*. Steele (1996) reported on the techniques he has developed for germinating seed in commercial quantities, and offers seed-propagated plants for sale. Several other commercial sources of seed-propagated plants are available, so there is hope that the threat to wild populations may diminish with time.

Epipactis Zinn

Catalogus Plantarum Horti Academici et Agri Gottingensis: 85. 1757.
Etymology: According to Correll (1950), the genus name is derived from a classical name used by Theophrastus for a plant used to curdle milk.

Epipactis (ep-i-pak'-tis) is a worldwide genus of approximately 25 species. The plants have lanceolate to broadly ovate plicate leaves that alternate on the stem. The underground rhizomes often form dense clumps. Most are native to Europe and Asia, but 2 species of *Epipactis* occur in the United States. Only one grows in Arizona; both grow in New Mexico.

Credit for describing the genus *Epipactis* has been given to multiple authorities over the years. Ames (1924) gave credit to Boehmer. Abrams (1940) accredited it to L. C. Richard. Correll (1950) and Luer (1975) among others attributed it to Swartz. L. O. Williams (1937), Hawkes (1965), Brackley (1985), and Stewart (1996) recognized Zinn as providing the original description. Brackley (1985) pointed out that Zinn validly published the genus name even though he did not use binomials for the species he described.

Key to the Species of *Epipactis*

1. Lip deeply three-lobed, the epichile elongated; found in wet places .... *E. gigantea*
1a. Lip not lobed, the epichile blunt, wider than long; found in dry places .... *E. helleborine*

Epipactis gigantea Hooker

Flora Boreali-Americana 2: 202. pl. 202. 1839.

Etymology: The specific epithet is from the Latin word for "gigantic," in reference to the relatively large size of the flowers and plants.

Synonymy:
Epipactis americana Lindley, Annals Magazine Natural History I. 4: 385. 1840.
Peramium giganteum (Douglas ex Hooker) Coulter, Contributions US National Herbarium 2: 424. 1894.
Serapias gigantea (Douglas ex Hooker) Eaton, Proceedings Biological Society of Washington 21: 67. 1908.
Helleborine gigantea (Douglas ex Hooker) Druce, Bulletin Torrey Botanical Club 36: 547. 1909.
Amesia gigantea (Douglas) Nelson & Macbride, Botanical Gazette 56: 472. 1913.

Common names: stream orchid, chatterbox, false lady's slipper, giant helleborine.

Plate 9

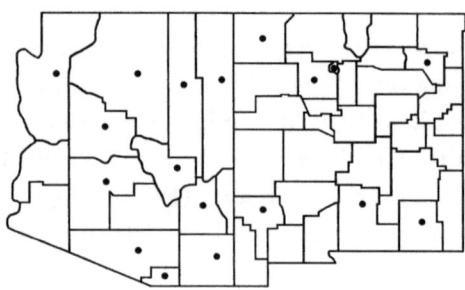

Map 8. Distribution of *Epipactis gigantea*

Description

Plant: glabrous, 20 to 100 cm tall.
Roots: underground rhizomes with a few thick fibrous roots.
Leaves: green, 5 to 11, narrowly ovate lanceolate to broadly ovate lanceolate, 1.5 × 7 cm to 11 × 25 cm; alternating on the stem; leaf size

decreases from bottom to top of stem, becoming bractlike near flowers.

Scape: 4 to 18 flowers, loosely scattered on top third of plant.

Floral bracts: lanceolate, large and leaflike on bottom flowers, to small on upper flowers; 2 × 7 cm to 0.3 × 1 cm.

Flowers: sessile, about 4 cm wide × 3 cm high; each with a bract.

Sepals: lateral sepals ovate lanceolate, slightly concave, dark green, slightly suffused with rose with faint darker veins, to 2.4 × 0.9 cm; dorsal sepal elongated ovate, slightly concave, dark green, slightly suffused with rose with faint darker veins, 2.0 × 0.8 cm.

Petals: ovate, oblique, concave, rose to intense pink, fading to green at apex, with pronounced darker veining, 1.7 × 0.7 cm.

Lip: 2.3 × 2.1 cm, constricted and hinged near the middle into basal hypochile and forward epichile; hypochile 1.8 × 1.5 cm, deeply three-lobed, lateral lobes with reddish raised veinlike ridges on a yellowish to greenish background; center lobe with raised, irregular papillae that are either reddish or yellowish; epichile 1.1 × 0.8 cm, mostly rose-colored with yellow markings, oblanceolate with two fleshy calli, fringe of the epichile extends beyond the calli.

Column: 1 cm tall with two spurs slightly below the stigma; anther cap 4 × 2 mm, light green, crescent-shaped; pollinia yellow powdery.

Capsule: ellipsoidal, pendent, 0.5 × 2.8 cm; forming on the lower flowers before the upper buds open.

Epipactis gigantea (jy-gan'-tee-ah) has a relatively large and colorful flower compared to some other orchids in the Southwest. The slightly concave sepals and petals are dark green, slightly suffused with rose and with darker veining, especially on the petals. The lip is constricted and hinged in the middle. The epichile vibrates with the slightest breeze, undoubtedly the characteristic that gave rise to the common name of chatterbox. The deeply three-lobed hypochile has a pouchlike appearance, which is why *E. gigantea* is sometimes called the false lady's slipper. The central lobe of the hypochile is covered with red or yellow papillae of irregular shapes. The epichile has two fleshy calli and additional rose and some yellow coloring. Two armlike projections protrude from the upper portion of the column slightly below the light green to yellow anther cap.

The papillae and sweet aroma of the flower apparently trick pollinators by mimicking their food supply. Ross (1988) reported that flies in the family Syrphidae are attracted to the flowers because their aroma mimics the honeydew smell given off by aphids. Syrphid flies normally lay their eggs in masses of aphids, which are the food supply for their larvae, and perhaps the papillae resemble aphids. However, there are usually no aphids on the orchids. The aroma of the flowers fools the fly into laying eggs among the supposed aphids, and in the process it picks up pollinia and deposits them on the next flower visited.

Distribution

Epipactis gigantea ranges from the Pacific coast eastward to Texas, Oklahoma, South Dakota, and Wyoming, and from northern Mexico to southern Canada. It is known from only 12 sites in Canada, where it is considered threatened (Brunton 1986) and may have been extirpated from several of these locations. It is widespread in Arizona, occurring in 11 counties. It is documented in every county except for La Paz, Pinal, Greenlee, and Yuma. Most likely it also grows in those counties, except possibly Yuma, because the proper habitats exist in each. From New Mexico, *E. gigantea* is known only from the 7 counties of Eddy, Grant, Harding, Los Alamos, Otero, San Juan, and Sandoval. Here, near its eastern limit, it should be expected to be less common than it is farther west. However, it is probably more common than the herbarium records indicate, and should be sought from more locations in New Mexico.

Habitat

Epipactis gigantea thrives in wet habitats at elevations between 1300 and 7120 feet (400 and 2140 meters) in Arizona and New Mexico. It grows at sea level and up to 8500 feet (2600 meters) in California. It grows in moist soil along banks of streams or in cracks in rocks of streambeds, with the water lapping against and sometimes over the roots. Large colonies, often with 1000 or more plants, may develop in these conditions. Some of the colonies are located such that they are several feet below the high water mark of peak flows resulting from storms. Most streamside colonies grow where covering trees provide protection from the sun, but some, especially in the north, grow in full sun at least part of the day. *Epipactis gigantea* is locally common along streams feeding the Colorado River in the Grand Canyon.

The stream orchid grows at the lowest elevation of any orchid in Arizona, extending down into riparian areas in desert canyons where it grows within view of many cacti such as prickly pear and cholla (*Opuntia* species), saguaro (*Carnegiea gigantea*), hedgehogs (*Echinocereus* species), and barrel (*Ferocactus* species). The orchids are usually under oaks on the canyon bottom, but palo verde and mesquite trees are on the canyon sides.

In the northeastern corner of Arizona, the most common habitat for the stream orchid is seeps in sandstone canyons. These seeps are easily recognized as damp green patches on the otherwise dry, barren sandstone walls. *Epipactis gigantea* grows in the seeps along with other riparian plants such as *Mimulus* species (monkey flowers), grasses, and sedges, in an environment known as *hanging gardens* because they are suspended above the canyon floor. Often it is impossible to approach the hanging gardens because of the steepness of the canyon walls, and the plants must be viewed with binoculars. The plants in the hanging gardens are usually smaller, 30 cm or less, and have fewer flowers than the ones that grow streamside or in seeps near canyon bottoms.

Blooming Season

Epipactis gigantea blooms between the very end of March and late June, although most flowering takes place in May and early June. The duration and quality of bloom depend on the yearly rainfall. If the precipitation is low, some seeps and normally wet creek bank areas partially dry up, and the orchids do not develop fully. Growth is blackened and stunted, and flowers either do not develop at all or abort without opening. Successive dry years result in the colonies shrinking.

Other orchids often associated with *E. gigantea* include *Platanthera sparsiflora* and *P. zothecina*. In some mountainous areas, *Hexalectris spicata, Cypripedium parviflorum, Corallorhiza maculata*, and *C. wisteriana* bloom nearby. *Goodyera oblongifolia* may bloom in the same area but later in the season.

Conservation

The stream orchid is distributed widely and exists in sufficient numbers across its range to be considered safe from most threats. However, in many areas of the Southwest, extensive groundwater pumping has reduced the flows of once permanent streams, and some colonies have disappeared or been reduced in size. Sivinski and Lightfoot (1995) cited historical populations in New Mexico that have been extirpated by water development at spring habitats. *Epipactis gigantea* is on List 2 in New Mexico, which contains plants that are rare because of limited distribution or low numerical density. It is not ranked as a rare plant in Arizona.

Epipactis helleborine (Linnaeus) Crantz

Stirpium Austriarum Fasciculus, ed. 2, Fasc. 6: 467. fig. 6. 1769.

Etymology: The Latin name means "like a hellebore," based on a supposed similarity to some plants in the buttercup family.

Synonymy:
Basionym: *Serapias helleborine* Linnaeus, Species Plantarum 2: 949. 1753.
Serapias latifolia (Linnaeus) Hudson, Flora Anglica: 393. 1762.
Epipactis latifolia (Linnaeus) Allioni, Flora Pedemontana 2: 152. 1785.
Epipactis helleborine (Linnaeus) Crantz var. *viridens* Gray, Botanical Gazette 4: 206. 1879.
Amesia latifolia Nelson & Macbride, Botanical Gazette 56: 472. 1913.

Common name: broad-leaved helleborine.

Plate 9

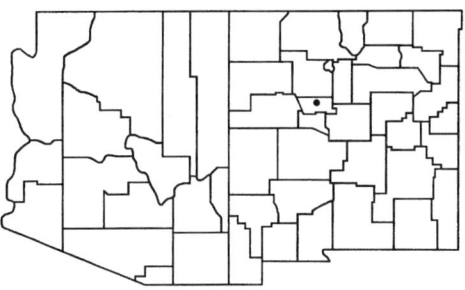

Map 9. Distribution of *Epipactis helleborine*

Description

Plant: height ranges from 15 to 100 cm, but more typically 30 to 50 cm; few to over 50 flowers.
Roots: thick mass of cordlike roots.
Leaves: four to seven, plicate, elliptic, 10 cm long × 5 cm wide.
Floral bracts: lanceolate, 3 cm long or more on lowest flowers, gradually reduced in length on upper flowers.

Flowers: greenish white, with rose to brown center on lip; approximately 1.5 × 1.2 cm.
Sepals: lateral sepals green, nearly ovate, 1.0 × 0.6 cm; dorsal sepal green, elliptic, acuminate, 1.1 × 0.4 cm.
Petals: whitish green, ovate elliptic, slightly cupped, 0.9 × 0.7 cm.
Lip: 1 cm long × 0.7 cm wide, without lateral lobes, constricted in the middle into two parts, basal hypochile and forward blunt epichile; hypochile cup- or saucer-shaped, with cup portion rose to dark brown and covered with nectar; epichile broadly triangular to nearly oblong, whitish green.
Column: greenish white, 6 mm high, yellow sticky pollinia protected by brownish anther cap.
Capsule: broadly elliptic to nearly spherical, about 1 cm long.

Epipactis helleborine (hel-le-bore'-een) can be identified by star-shaped flowers on plants that look somewhat like its close relative *E. gigantea*. However, even without flowers it is easy to tell the 2 species apart because compared to *E. gigantea*, the leaves of *E. helleborine* are generally flatter and wider and the elongating flower spike droops slightly near the tip. The star-shaped flowers are green and white in New Mexico, but in other states they often have a suffusion of pink in the petals. The sepals and petals are slightly concave and tilt forward, giving a cupped appearance to the flowers. The lip has a shiny pouch-like depression in the center of the hypochile that is dark brown and usually covered with droplets of nectar. The lip is hinged near the middle, though not as freely as in *E. gigantea*. The epichile is broadly triangular to ovate. The flowers have a pleasant aroma. Chlorophyll-free white and pink color forms of *E. helleborine* called forma *monotropoides* (Mousley 1927) occur in many parts of its range but are not known from this region.

Multiple species of wasps pollinate *E. helleborine* in other parts of its range (Judd 1971, Mousley 1927, Coleman 1995). Most likely, wasps are the pollination vectors here also, but pollination has not been observed to confirm that assumption.

Distribution

Epipactis helleborine is an Old World orchid, widely distributed in Europe, parts of Asia, and the northern tip of Africa. It is believed to have been introduced to the United States sometime before 1879, when it was discovered near Syracuse, New York. The introduction may have been deliberate by someone desiring plants to remind them of home, or the result of seeds contained in other material brought over from Europe. Once established, *E. helleborine* spread rapidly across the United States and Canada and is now in many states and provinces, coast to coast. It was first reported from New Mexico by Robert Sivinski, who is with the Forestry Division of the New Mexico Energy, Minerals, and Natural Resources Department. Sivinski found it in Bernalillo County and his is still the only confirmed location, although there are verbal reports of the plant in Santa Fe County. *Epipactis helleborine* has not yet been reported from Arizona, but should be expected there because of the abundance of suitable habitat and the freedom with which *E. helleborine* spreads.

Habitat

In Bernalillo County, *E. helleborine* grows in a mature cottonwood forest at approximately 5300 feet (1600 meters). The cottonwood forest is part of a larger plant community along the Rio Grande that contains meadows and mixed forests of willows, cottonwoods, and several established nonnative trees. The orchids grow in heavy duff from the cottonwoods, in partial shade to bright conditions. In other states *E. helleborine* grows in oak and pine forests. Since those habitats abound in Arizona and New Mexico, *E. helleborine* should be expected to show up in more diverse habitats now that it has a foothold in the region.

Blooming Season

Epipactis helleborine starts to bloom during the last week in June and remains in bloom through much of July. The blooming season should extend as more colonies become established in the Southwest, because flowering periods are much longer in areas where the plant is more numerous. Like many temperate zone terrestrial orchids, *E. helleborine* apparently does not need to appear above ground every year but can live off its mycorrhizal fungus for 1 year or more before emerging to bloom again. Light and MacConaill (1990, 1991) studied the population dynamics of *E. helleborine* over a 6-year period. Only two plants appeared every year, and most plants in the study appeared only once in the 6 years. Some plants appeared after a 3-year absence. Perhaps because of its limited habitat, no other orchids are known to grow in the same area as *E. helleborine*.

Conservation

Epipactis helleborine is extremely rare in New Mexico, with perhaps fewer than 100 plants in a single location. However, since it is an adventitious plant, *E. helleborine* should not be considered in the same category as *Hexalectris nitida* or *Piperia unalascensis*, native orchids also known from only a single area in this region. In at least one sense, *E. helleborine* is in a better situation than these other two species because it spreads readily and is capable of rapidly colonizing new areas.

Goodyera R. Brown

In Aiton, Hortus Kewensis, ed. 2, 5: 197. 1813.
Etymology: The genus was named in honor of John Goodyer, an English botanist.

The genus *Goodyera* (good-yer'-ah) consists of over 25 species distributed worldwide. Various authors list the number of species between 25 and 80. Four species occur in North America, and 2 of them, *G. oblongifolia* and *G. repens*, are in Arizona and New Mexico. All species produce a rosette of evergreen leaves marked with variable amounts of reticulation. All North American members of the genus share the common name of rattlesnake plantain.

Goodyera is in the subtribe Goodyerinae. Within this tribe several genera of small terrestrial orchids are collectively referred to as "jewel orchids." The jewel orchids are known for their beautiful foliage, and some species are cultivated for the foliage alone because the flowers are not showy. That does not apply to the 2 species in the Southwest, both of which have small but beautiful flowers.

Key to the Species of *Goodyera*

1. Plants usually shorter than 12 cm, flowers pure white, lateral sepals not recurved at apex   *G. repens*
1a. Plants usually taller than 20 cm, flowers with greenish or brownish sepals, lateral sepals recurved at apex   *G. oblongifolia*

Goodyera oblongifolia Rafinesque
Herbarium Rafinesquianum: 76. 1833.

Etymology: *Oblongifolia* is derived from Latin in reference to its oblong leaves.

Synonymy:
Spiranthes decipiens Hooker, Flora Boreali-Americana 2: 203. 1839.
Goodyera menziesii Lindley, Genera and Species Orchidaceous Plants: 492. 1840.
Peramium menziesii (Lindley) Morong, Memoirs Torrey Botanical Club 5: 124. 1894.
Peramium decipiens (Hooker) Piper, Contributions US National Herbarium 11: 208. 1906.
Epipactis decipiens Ames, Orchidaceae 2: 261. 1908.
Goodyera decipiens (Hooker) Hubbard, in Olmstead, Coville, and Kelsey, Standard Plant Names: 328. 1923.

Common name: Menzies' rattlesnake plantain.

Plate 10

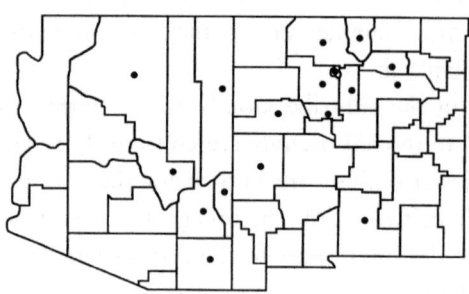

Map 10. Distribution of *Goodyera oblongifolia*

Description

Plant: evergreen, basal rosette of leaves, usually with white markings; one-sided inflorescence between 20 cm and 50 cm tall, pubescent, with over 30 small flowers.

Roots: creeping rhizome, usually branched, with fibrous roots.

Leaves: five to seven, evergreen in basal rosette, oblong elliptic, 3.5 × 8 cm, usually with white along central vein, and often with varying amounts of white veining on rest of leaf.

Floral bracts: lanceolate, 5 to 10 mm long.

Flowers: 1 cm long, 0.5 cm wide, and 0.5 cm high, greenish to white, pubescent on ovary and backside of sepals and petals; dorsal sepal and petals connivent, forming a hood over the column and lip.

Sepals: white with trace of green to greenish brown; dorsal sepal elliptic lanceolate, to 9 mm long × 4 mm wide; lateral sepals ovate, to 8 mm long × 4 mm wide, tips curl back tightly.

Petals: white, spatulate, oblique, to 9 × 4 mm.

Lip: white, saccate, with a narrow, linear, rounded, recurved apex.

Column: short, pointed, with pair of yellow pollinia.

Capsule: ovoid ellipsoidal, held in semierect position.

The plant habit and leaf pattern on *Goodyera oblongifolia* (ob-long-i-foe'-lee-a) provide for easy identification, even when the orchid is out of flower. The basal rosette of leaves, often in dense clusters, is usually marked with a distinctive white veining, or reticulation. Though most plants in the Southwest display at least some white along the center vein, reticulation varies from almost none, resulting in nearly solid green leaves, to very heavy over most of the leaf. Highly reticulated plants have been given the name *G. oblongifolia* var. *reticulata* Bovin, but plants matching that description are rare in Arizona and New Mexico. The leaves of *G. oblongifolia* persist for more than 1 year. The rosette of the leaves precedes flowering by 3 years (Sheehan 1992), and the leaves persist for at least 1 year after flowering. It is common to see this year's flowering growth accompanied by last year's dried flower stem held in the center of still fresh leaves.

The plants propagate by an underground creeping rhizome, developing multiple leads each year. As the center rosette dies off, these new leads set up new groups of rosettes, eventually leading to the massed colonies found in some areas. The plants also propagate by seed, producing rosettes in the fifth year after germination (Sheehan 1992).

The raceme is distinctly one-sided, with all the flowers facing more or less the same direction. The flowers have a trace of greenish brown in the sepals, the tips of which curl back tightly. The backsides of the sepals are covered with fine hairs. The petals are narrow at the base but dilate above the middle. The white lip is saccate, with a narrow, linear, rounded, recurved apex. On newly opened flowers, the sepals and petals remain clasped about the lip, with only a slight spreading at the tip. Some flowers retain this tight appearance until they fade. In others, as the flowers mature, the lower sepals fold back partway, and the relative positions of the column and lip change so the full beauty of the flowers can be seen. Even on the more open flowers, the dorsal sepal and petals remain connivent, forming a hood over the column and lip.

Ackerman (1975) identified the pollinators of *G. oblongifolia* as queen bumblebees of *Bombus occidentalis* Greene. Ackerman observed that the position of the column varies as the flower ages. On newly opened flowers, the column blocks access to the stigma but allows the bee to contact the sticky viscidium and remove pollinia. Newly opened flowers are at the top of the inflorescence, so removed pollinia are transported to another plant. The bees go to the lower flowers first, which have been opened longer and in which the column has moved, allowing pollinia on visiting bees to contact and adhere to the stigma. Ackerman reported an average rate of capsule set of 46.2 percent, with average rate for seed viability of 83.8 percent.

Distribution

Goodyera oblongifolia occurs in a large part of North America, from southern Alaska as far south as Mexico and as far east as the Gulf of St. Lawrence. In Arizona it is in the 6 counties of Apache, Cochise, Coconino, Gila, Graham, and Greenlee. Cochise County is the extreme southwestern limit of its range in the United States, but Luer (1975) reported it in portions of Mexico just south of Arizona. In New Mexico, Menzies' rattlesnake plantain occurs in the 11 counties of Bernalillo, Catron, Cibola, Los Alamos, Mora, Otero, Rio Arriba, San Miguel,

Sandoval, Santa Fe, and Taos. It also is most certainly in Colfax County because *G. repens* grows there, and *G. oblongifolia* is found in all other places where *G. repens* grows.

Habitat

Goodyera oblongifolia grows at elevations between 5745 and 10,000 feet (1750 and 3050 meters). Up to about 7000 feet, it grows among oaks. At the upper end of its elevation range, the most typical habitat for *G. oblongifolia* is humus of coniferous forests, in light to deep shade. It grows in forest duff on flat to steep terrain, and among rocks. In moist areas, *G. oblongifolia* will colonize and bloom on decaying logs. In the proper environment, colonies of hundreds of plants develop.

Goodyera oblongifolia can survive mild forest fires. In 1995, the Rattlesnake Canyon fire severely burned a portion of the Chiricahua Mountains, in Cochise County, Arizona, including parts of the habitat of *G. oblongifolia*. In 1997, plants were still growing inside both edges of the main burn area, where the fire was not as intense. However, in the central region of this particularly hot fire, no orchid plants remained.

Blooming Season

Because of its evergreen leaves, *Goodyera oblongifolia* can be found and identified any time of year, providing it is not covered with snow. The flower spikes start to elongate in June. At the lowest extreme of its elevation range, blooming often starts as early as the first week of July. At higher elevations, it is often still in bloom into the third week of September. In the middle elevations, it is reliably in bloom from the last week in July until the first week of September. Individual flowers of *G. oblongifolia* last for 2 weeks and seed capsules mature in 6 to 8 weeks.

Associated orchid species blooming at the same time may include *G. repens*, *Platanthera purpurascens*, and *Malaxis soulei*. Orchids in the same area but already out of bloom when *G. oblongifolia* opens include

Calypso bulbosa, all 4 *Corallorhiza* species, *Cypripedium parviflorum*, and *Schiedeella arizonica*.

Conservation

Goodyera oblongifolia is widespread and locally common. It grows in diverse habitats and is at least somewhat resistant to forest fires. Significant portions of its range in the Southwest are protected within wilderness areas and National Monuments. There are no apparent threats to its continued presence in the Southwest.

Goodyera repens (Linnaeus) R. Brown
In Aiton, Hortus Kewensis, ed. 2, 5: 198. 1813.

Etymology: *Repens* is from a Latin word meaning "creeping".

Synonymy:
Basionym: *Satyrium repens* Linnaeus, Species Plantarum 2: 945. 1753.
Peramium repens (Linnaeus) Salisbury, Transactions Horticultural Society of London I: 301. 1812.
Goodyera repens (Linnaeus) R. Brown var. *ophioides* Fernald, Rhodora 1: 6. 1899.
Peramium repens (Linnaeus) Salisbury var. *ophioides* (Fernald) Heller, Catalogue of North American Plants North of Mexico, ed. 2: 4. 1900.
Peramium ophioides (Fernald) Rydberg, in Britton, Manual of the Flora of the Northern States and Canada: 302. 1901.
Goodyera ophioides (Fernald) Rydberg, Brittonia I: 86. 1931.

Common names: lesser rattlesnake plantain, dwarf rattlesnake plantain, creeping goodyera, net leaf, squirrel ear, tessellated lesser rattlesnake plantain, white blotched rattlesnake plantain (var. *ophioides*), creeping ladies' tresses.

Plate 11

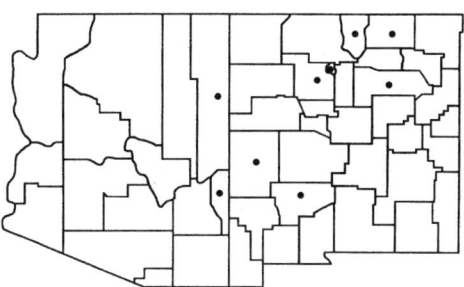

Map 11. Distribution of *Goodyera repens*

Description

Plant: about 9 to 12 cm tall (to 25 cm tall in Europe) with 15 to 22 flowers; leaves in basal rosette, rosette about 4.5 cm across.
Roots: creeping rhizome with scattered fibrous roots.

Leaves: evergreen, four to eight per rosette, ovate to oblong elliptic, 1.2 × 2.3 cm, solid dark green or very faintly marked with darker veining.

Floral bracts: lanceolate to elliptic lanceolate, 3 mm long, clasping ovary.

Flowers: white, about 3 × 3 mm, dorsal sepal and petals forming hood over column.

Sepals: white, back covered with fine white hairs; dorsal sepal broadly elliptical, concave, 4 × 2 mm; lateral sepals ovate, concave, 4 × 3 mm.

Petals: white, back covered with fine hairs, spatulate, 4 × 1 to 2 mm.

Lip: white, deeply saccate with narrow acute apex, 2 × 3 mm.

Column: tiny, with yellow pollinia.

Capsule: ellipsoidal.

Goodyera repens (re'-penz) is the smaller of the 2 *Goodyera* species in Arizona and New Mexico and is readily identified by its small stature and flower color. The lesser rattlesnake plantain is almost always less than 12 cm tall in bloom and is usually under 10 cm tall, while *G. oblongifolia* is at least 20 cm tall and usually much taller. *Goodyera repens* reportedly gets to a height of 25 cm in Europe and to nearly 20 cm in other parts of North America, but in Arizona and New Mexico, plants rarely exceed 12 cm. The fragrant flowers are a pristine white with only the slightest hint on some flowers of a green vein in the sepals. Dense white hairs on the sepals and petals add to the size of the diminutive flowers. The stem and ovary are also pubescent. The hairs are longer and more numerous than those of *G. oblongifolia*.

The dull, deep, almost bluish green leaves of *G. repens* more closely match the size of the rosettes of *Schiedeella arizonica* than they do the much larger rosettes of *G. oblongifolia*. Most of the rosette of four to eight leaves is close to the ground. On flowering plants, the uppermost leaf may be 1 to 2 cm up on the stem. Some leaves may display a faint marking of fine dark green, but more often the marking is missing. This unmarked pattern is typical of western plants. In the Northeast, leaves of *G. repens* are often delicately marked. Plants with extensive reticulation in the leaves have been called *G. repens* var. *ophioides*, which is not known from either Arizona or New Mexico.

Goodyera repens propagates by seed and vegetatively by long underground runners. Summerhayes (1968) reported that on European plants the runners develop underground for 5 years, lengthening each year, before reaching the surface and forming rosettes. It takes another 3 years for the rosette to bloom. Case (1987) believed that North American plants flower in a shorter period, but stated he could not verify that with documented observations. Even though it spreads by runners, *G. repens* does not form large colonies with hundreds of plants as does *G. oblongifolia*.

Smith (1993) reported that in Minnesota *G. repens* hybridizes readily with *G. tesselata*. In Arizona and New Mexico, *G. repens* and *G. oblongifolia* bloom in the same area and at the same time, but there are no obvious hybrids.

Distribution

Goodyera repens grows in temperate regions of the Northern Hemisphere. It is in many parts of Asia and mainland Europe, and Summerhayes (1968) reported it in Scotland. In North America, it ranges completely across Canada and into Alaska. In the United States it extends along the Appalachian Mountains into the Northeast, west to the Great Lakes and Rocky Mountains regions, and south into New Mexico and northeastern Arizona, where it is at its southern limit. It is known from only 2 counties in Arizona: Greenlee and Apache. Bennett, Johnson, and Kunzmann (1996) reported *G. repens* from Cochise County, but that claim is not supported by herbarium records and most likely is a misidentification of a small *G. oblongifolia*. *Goodyera repens* is more widely scattered in New Mexico, where it is found in the 7 counties of Catron, Colfax, Los Alamos, San Miguel, Sandoval, Sierra, and Taos.

Habitat

Goodyera repens grows in moderate to heavy shade in mixed fir, spruce, and aspen forest at elevations between 8000 and 10,000 feet (2438 and 3048 meters). The terrain it favors varies from nearly flat to fairly steep. The plants are thinly rooted in the duff and humus in damper parts of the forest. *Goodyera repens* requires more moist conditions than *G. oblongifolia*. A good indicator of the potential presence of *G. repens* is the amount of moss on the humus. Generally, the more terrestrial moss, the greater the moisture content of the humus, and the more likely *G. repens* will be present. It usually is found within a few yards of the bases of trees, rocks, or shrubs, where the shade helps maintain the necessary moisture content. *Goodyera repens* is so shallowly rooted that it sometimes colonizes moss-covered rocks, with its rhizome, roots, and runners visible at or just below the surface of the moss.

Blooming Season

Goodyera repens is one of the last orchids to bloom in mountain regions. Flower spikes emerge in late June and develop through the end of July. Flowering begins about the first of August and continues until the first part of September. The rosette of leaves fades after flowering, although the branching rhizomes have established offshoots that will bloom in the coming years.

Because it flowers late in the season, very few other orchids are in bloom when *G. repens* opens. Nearby, but generally in a slightly more xeric habitat, *G. oblongifolia* is also open, as is *Spiranthes romanzoffiana* in meadows and along streams above which *G. repens* grows. Earlier in the year, *Calypso bulbosa*, *Cypripedium parviflorum* var. *pubescens*, and *Platanthera purpurascens* bloom in or near the same habitat as *G. repens*.

Conservation

Goodyera repens is rare in Arizona, but significantly more common in New Mexico. In Arizona it is ranked G5/S3. The G5 ranking reflects its secure global position because of its worldwide distribution. The S3 ranking means it is uncommon or restricted, with between 20 and 50 occurrences in Arizona. In New Mexico, *G. repens* occurs often enough not to be ranked among its rare plants. In both states, suitable habitat is protected within designated wilderness areas.

Hexalectris Rafinesque

Neogenyton: 4. 1825.
Etymology: The genus name is derived from Greek words meaning "six" and "cock," in reference to the wavy crests on the lips that bear some resemblance to cock's combs.

Hexalectris (hex-a-lek′-tris) is a small terrestrial mycotrophic genus of six or seven members. As with *Corallorhiza*, there is little or no chlorophyll in the plants, and the leafless stems are shades of brown, tan, yellow, and purple. The genus was established by Rafinesque in 1825 and for many years was considered monotypic. Seventy-eight years later, Greenman (1903) added a second member, although his *H. mexicana* would later be reduced to synonymy with *H. grandiflora*, which had been described earlier in a different genus. Most of the other species were described in the 1940s. Members of the genus often are confused with *Corallorhiza*, and several *Hexalectris* species were originally described as coralroots. However, they are easily distinguished from *Corallorhiza* by the multiple raised crests down the center of the lip. All of the species and varieties of *Hexalectris* occurring in Arizona and New Mexico have either five, seven, or nine crests on the lip, even though the genus name suggests they have six.

The center of distribution for the genus is in Mexico, but *H. parviflora* is found as far south as Guatemala, and *H. spicata* is as far north as Maryland. Luer (1975) listed 5 species in the United States. Four species, one with two varieties, occur in Arizona and New Mexico.

Key to the Species of *Hexalectris*

1. Inflorescence shorter than 25 cm
    2. Sepals rolled back 360 degrees or more — *H. revoluta*
    2a. Sepals rolled back less than 180 degrees
        3. Lip with yellow lamellae — *H. warnockii*
        3a. Lip with reddish to purplish lamellae — *H. nitida*
1a. Inflorescence taller than 25 cm
    4. Column with rostellum
        5. Sepals rolled back less than 180 degrees — *H. spicata* var. *spicata*
        5a. Sepals rolled back 360 degrees or more — *H. revoluta*
    4a. Column without rostellum — *H. spicata* var. *arizonica*

Hexalectris nitida L. O. Williams
In Johnston, Journal Arnold Arboretum 25: 81. 1944.

Etymology: The epithet is from the Latin word meaning "shiny," referring to the appearance of the sepals and petals.

Synonymy: none.

Common names: Glass Mountain coralroot, shining cock's comb.

Plate 12

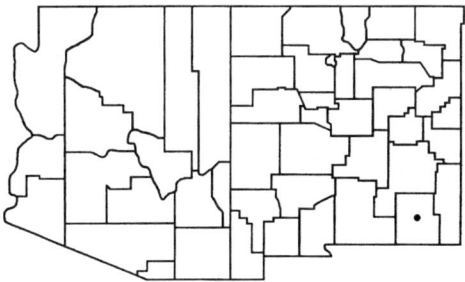

Map 12. Distribution of *Hexalectris nitida*

Description

Plant: mycotrophic, leafless, spicate with several bracts on stem, 13 to 20 cm tall, tan to dark brown with 12 to 20 waxy, shiny flowers.
Roots: none, grows from rough, thick rhizome.
Leaves: none, replaced by bracts on stem.
Floral bracts: lanceolate, 5 to 7 mm long.
Flowers: shiny, 1.2 × 1.2 cm, rose-brown to tan with white and purple on lip; sepals and petals spreading, turned back at apices.
Sepals: rose-brown with faint parallel veining; dorsal sepal elliptic lanceolate, 1.3 × 0.3 cm; lateral sepals oblanceolate with rounded apex, slightly falcate, 1.2 × 0.3 cm.
Petals: rose-brown with faint parallel veining, narrowly elliptic, 1.1 × 0.3 cm.

Lip: elliptic, three-lobed, 0.7 to 0.9 cm long × 0.4 to 0.5 cm wide; notch between lobes 1 to 1.4 mm deep; lateral lobes oblanceolate, white with reddish to purple markings; central lobe purplish, with seven raised ridges.

Column: 6 to 8 mm long, narrow, with minute wings at apex; four yellow pollinia.

Capsule: ellipsoidal.

Hexalectris nitida (nit′-i-da) can be identified at a glance by its very shiny flowers. The relatively short inflorescence bears a dense cluster of flowers that appear to be made of wax. Flower color is sufficient to separate it from *H. warnockii*, the only other *Hexalectris* of a similar size. On *H. nitida* the sepals and petals are a rich rose-brown, while on *H. warnockii* they are a reddish purple. Positive identification of *H. nitida* also can be based on characteristics of the lip. Its lip is less than 1 cm long, while all other *Hexalectris* species have lips longer than 1 cm. The shape of the lip is yet another diagnostic feature. On *H. nitida*, the flattened lip is shaped like an elongated arrowhead because the opposite edges of the lateral lobes are narrowly cuneate at the column. The flattened lips of all the other *Hexalectris* species in Arizona and New Mexico are much more rounded and are broadly cuneate at the column.

The lip of *H. nitida* invites careful study, and in the field this is best accomplished with a 10-power hand lens. The lateral lobes are cream white with purple markings along their juncture with the central lobe. The central lobe is mostly purple. Three raised ridges, or lamellae, start near the column and continue nearly the full length of the central lobe, stopping just short of the apex. Partway down the central lobe, within the portion joined by the lateral lobes, the first three ridges are joined by two others that also terminate just short of the apex. Two additional ridges start just beyond the notch where the lateral lobes diverge from the central lobe. These final two ridges are not as prominent or well formed as the others, but on the lower flowers they are clearly distinguishable, giving a total of seven ridges on the epichile of the lip. The lamellae have reddish to purplish markings.

Distribution

Hexalectris nitida is know from the Mexican state of Coahuila, and in the United States from Texas and New Mexico. In Texas it is widely scattered from Dallas to Big Bend National Park, and along its north-south border with New Mexico. In New Mexico it occurs in only Eddy County. *Hexalectris nitida* is not known from Arizona. Its western limit is in New Mexico and the adjacent parts of Texas just south of its location in New Mexico.

Habitat

The Chihuahuan Desert extends north from Mexico, across western Texas to its northern terminus in New Mexico. About 250 million years ago an ancient sea deposited an enormous limestone reef that today stretches from the Guadalupe Mountains of southwestern Texas to the city of Carlsbad, New Mexico. Within the boundaries of the fossil reef are Guadalupe Mountains National Park and Carlsbad Caverns National Park. Through millions of years, the natural forces of erosion have gradually eaten away at the limestone reef, carving out canyons in the rolling hill country and mountains. Some of the canyons are graced with permanent springs or year-round streams. Viewed from the canyon crest, the areas near the water sources stand out as bands of green in a mostly brown landscape. In the canyon bottoms near the water, the desert yields a little space, and the plant community changes from one dominated by various cacti, agave, and yucca, to stands of oak and juniper. The cacti, agave, and yucca occur in the understory of the oaks and junipers, but the shade from the trees and humus from decomposing leaves create a fragile environment capable of supporting orchids. Just out of the blazing sunlight, and protected somewhat from the heat, *H. nitida* grows on rocky canyon sides and bottoms at elevations between 4000 and 5000 feet (1219 and 1524 meters) in moderate to heavy shade from oaks and junipers. It is often near or under small shrubs and yuccas.

Blooming Season

Hexalectris nitida has a relatively short blooming season. The spikes first emerge in late May, and flowering begins by the end of June. Peak blooming occurs in July, but blooming may extend into early August. The number of flowering plants varies greatly from year to year, and individual plants may remain totally underground for more than 1 year.

Only a few other orchids share the desert canyons with *H. nitida*. *Hexalectris spicata* grows in the same habitat, but blooms a month or so earlier. Near springs or along streams, in slightly brighter settings it may be possible to find *Epipactis gigantea*. In Texas, *H. nitida* and *H. warnockii* bloom in the same woodlands, so *H. warnockii* may be in New Mexico also.

Conservation

Hexalectris nitida is extremely rare throughout its range, and especially so in New Mexico. It was first discovered in New Mexico in 1977 and is known from a single location in the Guadalupe Mountains of Eddy County. *Hexalectris nitida* is on List 1 in the Inventory of Rare and Endangered Plants of New Mexico. It is considered so rare in the state that unregulated collection could jeopardize its survival.

The site of *H. nitida* in New Mexico lies within Carlsbad Caverns National Park. Just across the state border in Culberson County, Texas, another site is within the boundaries of Guadalupe Mountains National Park. Therefore, its known locations in the region are protected from threats of development. The rangers in both parks are aware of the orchid's locations and of the need to protect the plants. Several oak- and juniper-lined canyons exist between Carlsbad Caverns National Park and Guadalupe Mountains National Park, and they should be searched for additional colonies of *H. nitida*.

Notes and Comments

Hexalectris nitida has not been observed in New Mexico since its original discovery. It was not found during searches by the author in 1997, 1998, and 1999. Repeated searches by park rangers and botanists have proved fruitless, so the possibility exists that *H. nitida* has been extirpated from the state. A slightly more optimistic view is to assume that since searches have not been made every year, the plant has just been missed owing to its habit of sporadic appearances.

Hexalectris revoluta Correll

Botanical Museum Leaflets 10: 19. 1941.

Etymology: The name *revoluta* refers to the revolute habit of the sepals and petals.

Synonymy: none.

Common names: curly coralroot, Correll's cock's comb.

Plates 12, 13

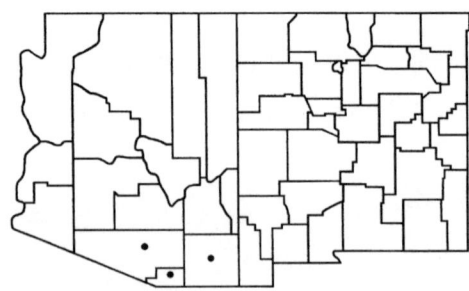

Map 13. Distribution of *Hexalectris revoluta*

Description

Plant: mycotrophic, leafless, spicate, pale pink to rose to tan, 40 to 50 cm tall, with 10 to 20 flowers.
Roots: none, grows from thick, coralloid rhizome.
Leaves: none, replaced by sheathing bracts on stem.
Floral bracts: ovate, 8 × 3 mm.
Flowers: rose-tan to whitish, 2.1 × 2.0 × 1.8 cm.
Sepals: pale rose-tan with light veining, revolute with outer third rolled back to form complete coil; dorsal sepal lanceolate, 2.2 × 0.8 cm; lateral sepals elliptic lanceolate, oblique, 2.0 × 0.8 cm.
Petals: pale rose-tan with light veining, elliptic to obovate, slightly falcate, 0.6 × 1.8 cm; revolute, with outer third rolled back in a full coil.
Lip: three-lobed, 1.5 × 1.2 cm; broadly elliptic in outline, white to pale rose-tan, with purple veining on lateral lobes, and purple raised ridges

on central lobe; lateral lobes oblong, with sinus about 2 mm; central lobe with five or seven raised ridges running entire length.

Column: narrow, curved; 1.5 cm high; white with purple shading at base; anther cap yellow, minute wings near the apex; eight yellow pollinia in four pairs.

Capsule: ellipsoidal, pendent.

Hexalectris revoluta (rev-o-loo'-ta) is identifiable on sight by the appearance of its sepals and petals. All three sepals and the lateral petals are rolled back along the outer third of their length more than 360 degrees to form a tight coil. This feature is not found on any of the other *Hexalectris* species in Arizona and New Mexico. *Hexalectris revoluta* may be initially confused with *H. spicata* with which it has several similarities, but the two are readily distinguishable. In addition to the revolute nature of its sepals and petals, *H. revoluta* can be differentiated from *H. spicata* by the shape of the lip and by its blooming season. The lateral lobes on *H. revoluta* are oblong, with the opposite sides of the lateral lobes essentially parallel to each other. On *H. spicata*, the lateral lobes are elliptic in outline, and the opposite sides are distinctly convex.

Hexalectris revoluta has a slender, sparsely flowered pale cream to tan leafless stem bearing four or five sheathing bracts. The sepals and petals, in addition to being revolute, are free and spreading. On *H. spicata*, *H. nitida*, and to some extent *H. warnockii*, the lateral sepals tilt forward and are held close to, if not against the column. The lobes of the lip are of intricate design. Their background color is whitish tan to rose-tan. The lateral lobes have distinct purple veining that is seen to be slightly raised if viewed under a microscope, but appears as simple lines to the unaided eye. The central lobe has five or seven raised purple ridges running its entire length, from near the column to the apex. The number of ridges varies from plant to plant and may even decrease on the upper flowers of the stem on large plants, but is always either five or seven. On particularly robust specimens, the outer ridges may split in two below the sinus with the lateral lobes, so that there are up to nine ridges on the lower portion of the central lobe.

Some plants of *H. revoluta* in Arizona differ from the typical flowers described above. They bloom in the same area and at the same time as the other plants, but on a shorter more sparsely flowered stem of richer tones. These plants are shorter than 25 cm and have five or six flowers. Compared to the larger form of *H. revoluta*, the sepals and petals are perhaps more revolute and reflexed backward to a greater extent, and the flowers have more purplish hues throughout. On two specimens examined in the field, the apex of the lip was solid purple, and the central lobe of the lip had only five ridges. The dorsal sepal was more lanceolate, and the lateral sepals and petals much more falcate. The column was identical to that of other *H. revoluta* orchids blooming nearby. These plants may represent a distinct, but as yet unnamed, variety of *H. revoluta*, or more conservatively, may simply demonstrate the inherent variability within the species.

Distribution

Hexalectris revoluta is known only from a few localities in northern Mexico, the Big Bend area of Texas, and Arizona. Within Arizona, it occurs in only four widely separated canyons in Cochise, Pima and Santa Cruz Counties, where it is at the northwestern limit of its range. It is not known from New Mexico.

Habitat

Though within the boundaries of the Sonoran Desert, parts of Cochise County, Pima County and adjacent Santa Cruz County consist of rolling hills, where with slight increases in elevation, the desert yields first to grasslands and then to woodlands. Canyons often cut through the terrain, particularly as the hills gain elevation nearer the mountains. With the increase in elevation, the mesquite-studded grasslands become mixed with juniper and oaks, even more so at the bottoms of the canyons. Some of these oak woodland canyons, at elevations between 4500 and 5200 feet (1370 and 1580 meters), are home to *H. revoluta*. Though oaks dominate, trees and shrubs in the canyon include juniper, mesquite, Arizona walnut, acacia, and desert willow. The canyons are seasonal water sources, but there may be a several-year interval between storms heavy enough to create running water. Even so, the orchids are protected from most of the force of the runoff. *Hexalectris revoluta* grows under the trees and shrubs on the edges of the canyon bottoms, and on hillsides leading up from the canyon. Under the oaks it is in heavy leaf litter, but closer to the canyon bottom *H. revoluta* is found in very thin humus layers. In some areas, the orchids are among rock outcrops or on the edges of rocky cliffs.

Blooming Season

The pale spikes of *H. revoluta* appear in April and the blooming season lasts from mid-May to the middle of June. The quality of bloom is unpredictable but may be influenced by rainfall in the previous or current year. Some years all the plants that send up spikes will put on a good display of flowers. Other years, none of the plants that sprout in an area complete blooming. The flowers may blacken, shrivel, or be eaten. The number of plants appearing each year is variable. In one study area, the number of plants observed varied from 1 in 1996, to 18 in 1997, to 9 in 1998, to 4 in 1999. None of the plants that sprouted in 1998 managed to produce open flowers.

The harsh environment of *H. revoluta* is not shared by many other orchids. *Hexalectris spicata* var. *arizonica* grows in the same habitat, but when the flowers of *H. revoluta* are open, the richer pink spikes of *H. spicata* var. *arizonica* are barely above ground. An occasional *Malaxis soulei* may grace the more densely wooded parts of the habitat of *H. revoluta*, but they will not appear until after the monsoons start.

Conservation

Hexalectris revoluta is extremely rare throughout its range and should be nominated for federal consideration as an endangered species. Some of its habitat in Arizona is at extreme risk from mining development. One of its major locations was briefly part of a planned land exchange between the U.S. Forest Service and a mining company until falling copper prices forced postponement of the deal. It has not been observed in one of the four known locations since 1981, despite repeated recent searches. *Hexalectris revoluta* is a recent addition to the flora of Arizona and is not yet ranked by the state. However, when ranked, it should be classified G1/S1 due to extreme rarity throughout its range.

Notes and Comments

Hexalectris revoluta was first observed in Arizona in 1981 by L. Toolin and F. W. Reichenbacher, who found it in a wooded canyon in south central Pima County on lands belonging to the Tohono O'odham Indian Nation. In 1986, S. McLaughlin found it on National Forest lands in eastern Pima County. All three of these individuals believed it to be *H. spicata* based on keys in the various floras of Arizona. In 1996 McLaughlin led the author to the site of his discovery, and they found one plant in spike. When it later flowered, the author recognized it as *H. revoluta* and documented it as an addition to the flora of Arizona (Coleman 1999).

Hexalectris spicata (Walter) Barnhart
Torreya 4: 121. 1904.

Etymology: The specific epithet refers to the spicate inflorescence.

Synonymy:
Basionym: *Arethusa spicata* Walter, Flora Caroliniana: 222. 1788.
Corallorhiza arizonica S. Watson, Proceedings American Academy of Arts and Sciences 17: 379. 1882.
Corallorhiza spicata (Walter) Tidestrom, in Tidestrom and Kittell, Flora of Arizona and New Mexico: 733. 1941.

Common names: crested coralroot, cock's comb, dragon's claw.

Plates 13, 14

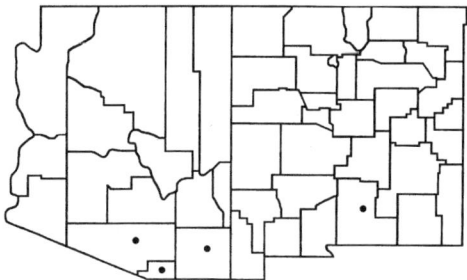

Map 14. Distribution of *Hexalectris spicata* var. *arizonica*

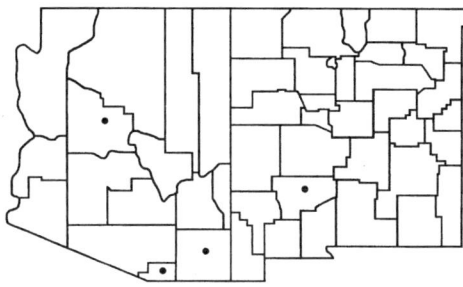

Map 15. Distribution of *Hexalectris spicata* var. *spicata*

Description

Plant: mycotrophic, leafless, spicate, 27 to 65 cm tall, nearly 1 cm thick at base; 9 to 20 flowers, dark pink to tan to dark brown.

Roots: branched, circumferentially ridged rhizome with few if any fibrous roots.

Leaves: none, reduced to bracts on stem, up to five bracts sheathing stem.

Floral bracts: lanceolate to oblanceolate, 0.8 cm long.

Flowers: 1.5 to 2.5 cm wide × 2.0 to 2.5 cm high; tan to dark brown, with purple ridges and stripes on lip.

Sepals: tan to brown with faint veining; dorsal sepal lanceolate to elliptic lanceolate, 0.8 cm wide × 1.8 cm long; lateral sepals oblong elliptic, slightly falcate, 0.7 cm wide × 1.7 cm long.

Petals: tan to brown with faint veining, oblanceolate, slightly falcate, 0.6 cm wide × 2.0 cm long.

Lip: white with purple dots, stripes, and ridges; three-lobed, 1.4 cm wide × 1.6 cm long, five or seven raised purple ridges down central lobe with purple lines in lateral lobes.

Column: white, 1.3 cm high × 0.3 cm wide, slight hour-glass shape, curved when viewed from side; pollinia yellow.

Capsule: ellipsoidal, 1.2 cm long × 0.6 cm in diameter.

Hexalectris spicata (spi-ka'-ta) appears above ground only to flower as a stout, leafless, spicate inflorescence. Since these mycotrophic plants lack significant amounts of chlorophyll, the flower stalks are shades of brown, ranging from a yellowish tan to rich, almost reddish dark brown. The sepals and petals are a light rose-tan, with darker reddish brown stripes, somewhat reminiscent of the stripes in *Corallorhiza striata*. The sepals spread widely and are slightly recurved at the tips. The petals are also recurved at the tips and bend forward over the column. Though definitely recurved, the sepals and petals roll back less than 90 degrees. The three-lobed lip is white, with five raised crests running the entire length of the central lobe. The crests are reddish purple, and the lateral lobes have reddish purple veining. The apex of the lateral lobe does not uncurl fully on plants in this area, giving a somewhat cupped appearance to the forward portion of the lip.

Hexalectris spicata can be distinguished easily from *C. striata*, with which it is sometimes confused, by the three-lobed shape of the lip and the raised ridges down the central lobe. The lip of *C. striata* is entire, with fused lamellae on only about the basal third. *Hexalectris spicata* is closely related to *H. revoluta*, which shares the same habitat. They are most easily differentiated by the habit of the sepals and petals. On *H. spicata* the sepals are recurved, but less than 90 degrees. The petals on *H. spicata* usually lean forward over the column and are only slightly, if at all, recurved. The sepals and petals of *H. revoluta* recurve more than 360 degrees, forming tight circles at the apex.

A second variety of *H. spicata* grows in Arizona and New Mexico. Catling and Engel (1993) described a self-pollinating variety they named *H. spicata* var. *arizonica*. The principle difference between the varieties is the lack of a rostellum in *H. spicata* var. *arizonica*. Since the function of the rostellum is to separate the pollen from the stigma, lack of a rostellum enables autogamy. The crests on the lip are generally shorter in *H. spicata* var. *arizonica*, although there is some overlap in crest height among the varieties, so it is not always useful as a diagnostic character. Normally the flowers on *H. spicata* var. *arizonica* either do not open at all or open only partially. However, sometimes, though only rarely, one or more flowers on a stem will open completely. Separating *H. spicata* var. *arizonica* from *H. spicata* var. *spicata* based on flower characteristics then requires inspecting the column. Even on freshly opened flowers, the yellow pollinia on *H. spicata* var. *arizonica* will have slipped out from under the anther cap and onto the stigma. This slippage also occurs on flowers that never open, effecting self-pollination.

Hexalectris spicata var. *arizonica* can be identified fairly accurately while still in early spike. The spikes have a pinkish brown cast, while *H. spicata* var. *spicata* is much more tan to brownish.

Distribution

Hexalectris spicata var. *spicata* is widely scattered in the United States from Florida to Maryland and westward to Arizona and New Mexico. It also occurs in much of northern Mexico. In Arizona, *H. spicata* var. *spicata* occurs in just 3 counties. It has long been known in the counties of Cochise and Santa Cruz. In 1992 Marc Baker and Theresa Wright made a major range extension when they found an extended colony of *H. spicata* var. *spicata* in Yavapai County, where it is at its western limit. In New Mexico, *H. spicata* var. *spicata* occurs in Sierra County.

Hexalectris spicata var. *arizonica* grows at the western limits of the taxon in Texas, New Mexico, Arizona, and parts of Mexico. In Arizona it grows in the counties of Pima, Cochise, and Santa Cruz, and in New Mexico it occurs in Otero County.

Habitat

Both varieties of *H. spicata* share a similar habitat at elevations between 5400 and 6500 feet (1650 and 1980 meters). At the lower end of their elevation range they grow in oak woodlands, on the wooded sides of canyons, and on canyon bottoms. At the upper end of their elevation range, the forest cover changes to mixed oak and conifers. The orchids typically are under the drip line of the oaks, pines, and companion shrubs, where they are often well hidden from casual observation. Rarely are the plants out in the open. The rhizomes are 10 to 15 cm (4 to 6 inches) deep in humus to very rocky soil. The plants are adaptable to a wide range of lighting conditions, from direct sun much of the day to deep shade.

Blooming Season

The varieties of *H. spicata* have different blooming seasons. *Hexalectris spicata* var. *spicata* blooms from about mid-June to early July from spikes that sprout in early May. The thick, emerging flower stem has multiple sheathing bracts and looks like an enormous coralroot. *Hexalectris spicata* var. *arizonica* produces flowers from late July to late August. Were it not for the rare open flower, it almost would be incorrect to say it blooms since the self-pollinating flowers usually do not fully open. The plants of both varieties seldom flower 2 years in a row and often will skip several years before reappearing. Not all plants that emerge with spikes and buds proceed to flowering. In both varieties a significant percentage (10 to 80 percent, depending on the year) of spikes and buds abort without fully developing, and others are browsed upon. Some years in a colony of perhaps 50 to 60 plants that initiate spikes, only 1 or 2 complete the blooming cycle.

Other orchids that bloom in the same habitat as *H. spicata* include *H. revoluta*, *H. warnockii*, *Corallorhiza wisteriana*, *Malaxis corymbosa*, and *M. soulei*. *Epipactis gigantea* may occur in wet areas nearby.

Conservation

Hexalectris spicata is relatively rare in Arizona and is ranked G5/S3S4. The rules of ranking apply to the species as a whole, so *H. spicata* var. *spicata* and var. *arizonica* are included together. The G5 refers to its secure global ranking because of the total number of occurrences and widespread distribution. The S3S4 ranking within Arizona indicates there are between 51 and 100 occurrences. In New Mexico, *H. spicata* is a List 1 plant, which means it is considered rare and endangered. *Hexalectris spicata* var. *spicata* is less common than *H. spicata* var. *arizonica*. It grows in small colonies, and only a few plants bloom each year. *Hexalectris spicata* var. *arizonica* is slightly more common but still rare. It usually is found as widely scattered individuals, though some small colonies develop up to a half-dozen plants.

Much of the habitat of *H. spicata* is at risk to mining activities. Some plants of both varieties grow within the Chiricahua National Monument, and that location is safe from development.

Hexalectris warnockii Ames and Correll
Botanical Museum Leaflets 11 (1): 8. 1943.

Etymology: The species was named in honor of Barton H. Warnock, a student of the flora of the Glass and Chisos Mountains in Texas.

Synonymy: none.

Common names: purple-spike coralroot, Texas purple-spike.

Plate 14

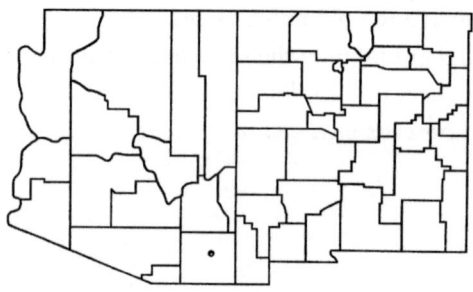

Map 16. Distribution of *Hexalectris warnockii*

Description

Plant: mycotrophic, leafless, slender spicate purple spike between 7 and 23 cm tall, with one to six flowers.
Roots: none, grows from coralloid rhizome.
Leaves: none, replaced by bracts on stem.
Floral bracts: small, purple.
Flowers: about 2.2 cm wide × 3.5 cm high, reddish purple with white and yellow on lip; sepals spreading, petals spreading or held forward next to column.
Sepals: reddish purple; dorsal sepal linear oblong, 1.9 × 0.3 cm; lateral sepals three-nerved, linear falcate, 1.7 × 0.4 cm.
Petals: falcate, wider at the apex than at the base, 1.6 × 0.3 cm.

Lip: three-lobed, 0.6 cm wide × 1.3 cm long; lateral lobes mostly purple from heavy veining; central lobe white with five ridges crested with yellow down the center.

Column: white with purple highlights, especially around the apex, curved, 1.2 cm tall; anther cap, purple; eight pollinia in four pairs.

Capsule: elongated ellipsoidal, almost cylindrical.

Hexalectris warnockii (wahr-nock'-ee-eye) is one of the most difficult of the southwestern orchids to see, even if you know where to look. The red-purple of its leafless spike blends with the shades of brown common to the oak litter of the forests where it grows. The sepals and petals are the same red-purple as the spike. The lip, a combination of white, yellow, and purple, adds delightfully contrasting colors to the arrangement and aids somewhat in finding the plants. The lip is three-lobed and mostly white. The lateral lobes, so heavily veined as to appear almost purple, curl upward to wrap around the column. The central lobe is white, with five raised ridges along its length. The ridges are white at their base but change to a bright yellow at their crests. The lip spreads out at the tip and has a purple spot at the apex, and wavy margins. The raised ridges are straight and parallel for two-thirds of their length but become wavy on the lower portion of the lobe.

The lateral sepals are free and spreading, and on some plants the petals are also, resulting in a very open flower. On other plants, the petals tilt forward along the column, combining with the lateral lobes of the lip to totally conceal the column. The flowers do not have any noticeable aroma. Except for the few plants that set fruit, most blacken and wither shortly after flowering. On those that set fruit, the stems and fruit lose some of their purple luster as the capsules mature, becoming more tan or yellowish.

Distribution

In the United States *H. warnockii* occurs in a few widely scattered sites in the states of Texas and Arizona. It is widely scattered in Texas from Dallas to the Big Bend area. It was discovered in Baja California, Mexico, fairly recently (Salazar 1991), extending its range far to the west. In Arizona it grows only in Cochise County, which is its northern limit.

Habitat

The Texas purple-spike grows in mixed oak woodland at elevations between 5000 and 6000 feet (1524 and 1829 meters). The forest cover is mostly silverleaf oak with some pine, madrone, and manzanita. The tiny orchids occur in moderate to heavy shade. The plants grow near or among rocks or near fallen logs, as if seeking moisture. Some plants grow within a yard or two of prickly pear and cholla cacti, and others hide under the branches of trees and shrubs.

Blooming Season

Though late-summer bloomers, the yearly appearance of *H. warnockii* is not as closely keyed to the start of the monsoon rains as that of other summer-blooming species such as *Malaxis*. The first bright pink-purple spikes emerge in late July to early August, varying only about a week depending on whether the rains come early or late. The spikes darken to a duller red-purple by the time the first flowers open in early to mid-August. Individual flowers remain open for only a day or two. However, plants appear over a period of several weeks, so the total blooming season is about 3 weeks long and may last until the first week of September. Many of the plants that initiate flower spikes and buds do not develop to flowering. In 1995 and 1996, flowers developed from less than 15 percent of the spikes. In 1997 only four flowers opened out of over 60 plants that appeared. On the others, stems, buds, and ovaries withered before anthesis. It appears as if some spikes shrivel to the ground from lack of moisture or from the extreme heat of the Arizona summers. Others remain standing but turn black as if destroyed by mold or fungus. Even on plants that open one or more flowers, adjacent buds may shrivel without opening. Failure of all flowers on a spike to develop is so common that plants having all buds open, instead of having some shriveled, are rare. Fruit set is also extremely rare. In 1996, out of almost 80 plants that started spikes, a potential of over 300 flowers, fewer than 40 flowers eventually opened, and only 2 set capsules. Individual plants usually do not bloom in successive years and may not reappear for several years, remaining totally underground until ready to bloom again.

Hexalectris warnockii blooms in the same area as *H. spicata*, *M. soulei*, and *M. corymbosa*. *Hexalectris spicata* is long since in capsule, but the *Malaxis* species bloom the same time as the Texas purple-spike.

Conservation

Hexalectris warnockii is rare throughout its range and should be a candidate for federal listing as an endangered plant. It is ranked G2/S1 in Arizona, meaning it has fewer than 20 occurrences across its range and fewer than 5 in the state. Within Arizona it is confirmed from only two localities based on herbarium specimens. Wentworth (1982) placed it in a third location in the Mule Mountains. Fortunately, one of its confirmed locations is within a national monument, and therefore its habitat there is protected. At this site it appears regularly each year. The plants were observed only once at the second confirmed location, in 1992, and were not seen during subsequent searches from 1995 through 1999. Suitable habitat for *H. warnockii* exists in many of the canyons of southeastern Arizona, but the orchid has not been located in any but the just mentioned localities.

Notes and Comments

The author found his first plants of *Hexalectris warnockii* in 1995, blooming in the Chiricahua National Monument. The search area was based on information from the late Hiram Parent, who had photographed it in 1963. Much to his delight, a prime specimen was in bloom. When the location was described to Mr. Parent, he surmised it was most likely the same plant he had seen over 30 years earlier!

Listera R. Brown

In Aiton, Hortus Kewensis, ed. 2, 5: 201. 1813.
Etymology: The genus was named in honor of Dr. Martin Lister, an English botanist and physician.

Listera (lis-ter-ah) is a circumboreal genus of 25 species. Most species are less than 20 cm tall, with numerous tiny green or brown flowers. Plants in this genus are commonly referred to as *twayblades* because of the pair of opposing leaves midway on the stem. The size and appearance of *Listera* species are very similar to the size and appearance of some species of *Malaxis*, and in Pima County, Arizona, species from both genera grow in the same area and bloom at the same time. However, it is very easy even for the novice to separate the two, at least in Arizona and New Mexico. *Listera* species can be readily identified by the presence of two leaves, while all *Malaxis* species in Arizona and New Mexico have a single leaf.

Nieuwland (1913) pointed out that the genus names *Listeria* Necker and *Bifolium* (Lobelius) Petiver-Millan predated *Listera*. However, Correll (1950) conserved the name *Listera*.

There are 8 species of *Listera* native to the United States and Canada. *Listera convallarioides* occurs only in Arizona, and *L. cordata* occurs only in New Mexico.

Key to the Species of *Listera*

1. Lip bilobed, deeply forked — *L. cordata*
1a. Lip minutely trilobed, not forked — *L. convallarioides*

Listera convallarioides (Swartz) Nuttall
Genera North American Plants 2: 191. 1818.

Etymology: *Convallarioides* means "like convallaria" or "like lily-of-the-valley," referring to a supposed resemblance of the orchid to the lily-of-the-valley.

Synonymy:
Basionym: *Epipactis convallarioides* Swartz, Vetenskaps Academieus Handlingar Stockholm 21: 232. 1800.
Ophrys convallarioides (Swartz) Wright ex House, Bulletin Torrey Botanical Club 32: 380. 1905.
Bifolium convallarioides (Swartz) Nieuwland, American Midland Naturalist 3: 129. 1913.

Common name: broad-leaved twayblade.

Plate 15

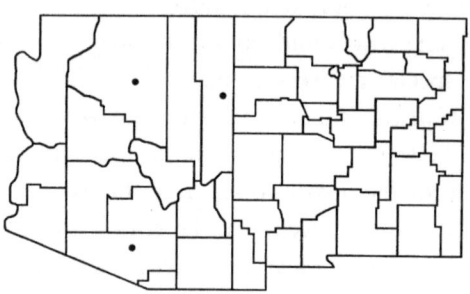

Map 17. Distribution of *Listera convallarioides*

Description

Plant: between 10 and 20 cm tall and with terminal raceme; two opposite leaves, midway on stem; stem glabrous below leaves, covered with fine hairs above leaves; one small bract between leaves and flowers.
Roots: three to eight, fibrous, 1 to 4 cm long.
Leaves: oblong to round, each about 4.5 × 3.5 cm.
Floral bracts: lanceolate, approximately 0.3 mm long.

Flowers: pale green, up to 17 per plant, each approximately 0.4 × 0.8 cm.

Sepals: dorsal sepal ovate lanceolate to nearly elliptic, with slightly darker green stripe down middle; lateral sepals lanceolate, slightly falcate, with slightly darker green stripe down middle; back of sepals covered with fine hairs.

Petals: acute clavate, slightly falcate, 3.7 × 0.7 mm; faint darker green stripe down center, back covered with fine hairs.

Lip: translucent, stout fiddle shape, 1.0 cm long × 0.5 cm wide at the broadest spot near the tip; slightly notched to barely trilobed at the apex; contracting abruptly as it nears the column, forming a narrow claw about 1 mm wide; slightly raised ridge down center of lip from near claw, to near apex, more prominent at claw.

Column: curved, 4 mm long, with two elongated yellow pollinia.

Capsule: ellipsoidal to nearly spherical, 0.2 × 0.6 cm.

The flowers of *Listera convallarioides* (kon-val-er-ree-oy'-deez) are reportedly sometimes marked with purple (Bingham 1939), but those in Arizona are solid green. The lip is the most prominent feature of the flower, and its shape makes it easy to recognize this species. The lip is translucent and narrows down abruptly as it nears the column, forming a claw. The apex is minutely three-lobed but often appears merely notched. The diminutive lanceolate sepals and linear petals are sharply reflexed around the ovary, and the column curves prominently over the lip.

Ramsey (1950) reported that *L. convallarioides* employs a triggered rostellum to aid pollination. When a visiting insect bumps a minute projection from the rostellum, the pollen masses are fired onto the intruder, which then transports them to the next flower visited.

Distribution

Listera convallarioides is widely distributed in North America, from southern California into Canada and east to New England. A finger of distribution heads south along the Rocky Mountains, with a slightly disjunct presence in Arizona. Within Arizona it grows in only three widely separated areas in Pima, Coconino, and Apache Counties. Its occurrence in Pima County was known since 1907 and represents the southern limit for this species. It was not until 1986 that it was discovered along the Mogollon Rim in Coconino County, and not until 1997 was it documented in Apache County. It does not occur in New Mexico.

Habitat

Listera convallarioides grows in constantly moist locations in pine and fir forest at elevations between 7000 and 8600 feet (2100 and 2600 meters). In Arizona it grows along the banks of perennial streams or seeps, in mosses, or directly in the damp soil, often hidden under grasses, ferns, or bushes. Relatively diminutive plants with tiny inconspicuous flowers create a challenge for those searching for it. In Apache County, Arizona, the plants are found most easily early in the spring while they are not yet in bloom. By the time the plants bloom in late summer, ferns that were just unfurling in spring are now at their full height and totally obscure the twayblades.

Blooming Season

Listera convallarioides blooms from early July to mid-August. Because it grows along permanent streams, it is not subject to variation in blooming season based on timing of the monsoon season. However, the quality of bloom in Pima County, Arizona, is often influenced by the strength of the summer rains. The stream where it grows can experience heavy runoff from thunderstorms, and during peak flows the plants may be briefly submerged and the flowers battered by the raging waters. Though such strong storms occur every summer, they do not always occur in the drainage system for the *Listera*, and in many years the flowers open and bloom normally without being damaged by high water.

Other orchids that may bloom near *L. convallarioides* include several *Platanthera* species, *Malaxis abieticola*, and *M. soulei*. Growing in the same area but already out of bloom when the twayblades open are *Corallorhiza maculata*, *C. striata*, *C. wisteriana*, and *Schiedeella arizonica*. In Apache County, *Goodyera oblongifolia* and *G. repens* grow nearby but bloom later.

Conservation

Listera convallarioides is common throughout much of its range but is relatively rare in Arizona. It has been found in three stream systems on Mt. Lemmon in Pima County, but only one small colony has been verified since 1914, where it was still growing when last visited in 1999. It is slightly more common along the Mogollon Rim where it occurs in four drainages. There is only one known colony in Apache County. The wide separation of the populations probably means *L. convallarioides* is not endangered in Arizona, although the remaining plants in Pima County would seem to be at risk if a major flood occurs in the canyon where they grow. *Listera convallarioides* is ranked G5/S1, which means it is secure across its range but has fewer than six occurrences in the state.

Listera cordata (Linnaeus) Brown

In Aiton, Hortus Kewensis, ed. 2, 5: 201. 1813.

Etymology: *Cordata* derives from Latin, meaning "heart-shaped," in reference to the shape of the leaves.

Synonymy:
Basionym: *Ophrys cordata* Linnaeus, Species Plantarum: 946. 1753.
Epipactis convallarioides Bigelow, Florula Bostoniensis, ed. 2: 323. 1824.
Listera nephrophylla Rydberg, Memoirs New York Botanical Garden 1: 108. 1900.
Ophrys nephrophylla Rydberg, Bulletin Torrey Botanical Club 32: 610. 1905.
Bifolium cordatum Nieuwland, American Midland Naturalist 3: 129. 1913.
Listera cordata (Linnaeus) R. Brown var. *nephrophylla* (Rydberg) Hultén, Flora Aleutian Islands: 145. 1937.

Common name: heart-leaved twayblade.

Plate 15

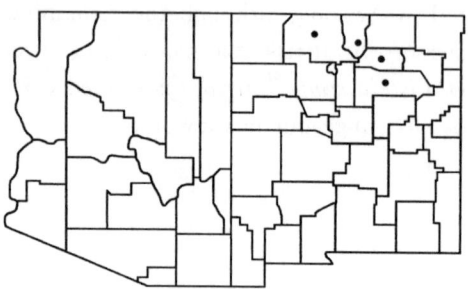

Map 18. Distribution of *Listera cordata*

Description

Plant: to 25 cm tall elsewhere but usually not much more than 10 cm in this area; two heart-shaped opposite leaves midway on the stem; up to 30 flowers.

Roots: 10 to 15 stringlike roots about 5 cm long from base of swollen stem.

Leaves: cordate, about 2 × 2 cm.

Floral bracts: tiny, 1 mm long.

Flower: green (green to reddish elsewhere), perianth star-shaped, with protruding forked lip, about 5 mm wide × 8 mm long.

Sepals: green; dorsal sepal elliptical, slightly cupped, 2 × 1.3 mm; lateral sepals lanceolate, 2 × 1 mm.

Petals: green, broadly elliptical, 2 × 1.5 mm.

Lip: green, linear over first half, deeply forked for lower half, 2 mm wide × 4 mm long, two hornlike appendages near column.

Column: short, stubby, with yellow granular pollinia in linear bundles.

Capsule: nearly spherical.

Listera cordata (kor-dah'-ta) is very easy to distinguish from *L. convallarioides* because the shape of the lip is diagnostic. The lip of *L. cordata* is deeply forked for half of its length, while the lip of *L. convallarioides* is shallowly trilobed at the apex. The plant structure in the two species is very similar, with two opposite leaves midway on the stem. Very rarely *L. cordata* will have three leaves instead of the usual two, with the third leaf between the usual pair and the lowest flower, or will have the two leaves alternate instead of opposite. The flower of *L. cordata* is whimsical. The deeply forked and spreading lip resembles legs, and two hornlike projections near the base of the lip resemble arms, giving the flower an elflike appearance. The elfin appearance is complete if you imagine that the petals and sepals spreading like a star over the lip are the hat.

Luer (1975) and Correll (1950) recognized two varieties of *L. cordata* in the United States, differentiating between *L. cordata* var. *cordata* and *L. cordata* var. *nephrophylla* (Rydberg) Hultén based on flower color and flower and leaf size. They maintained that the flowers of *L. cordata* var. *nephrophylla* are always green and the flowers and leaves are slightly larger than the flowers of *L. cordata* var. *cordata*, which vary from green to red. Other authors such as Hitchcock (1969) and Calder and Taylor (1968) did not recognize the two varieties, stating that flowers of intermediate colors make separation of the taxa on the basis of flower color not feasible. Even though only plants with green flowers have been identified in New Mexico, based on total knowledge of the species, the merged approach advocated by Hitchcock and Cadler and Taylor is followed in this work.

Ackerman and Mesler (1979) described the pollination mechanics of *L. cordata*. Three pressure-sensitive hairs at the tip of the rostellum act as a trigger for pollination. A droplet of glue is ejected onto a visiting insect that touches one of the hairs, then the pollinia are released to fall on the glue. Once the pollinia have been removed, the spread rostellum remains as a protective cover of the stigma for approximately 1 day, after which it gradually lifts, exposing the stigma. Visiting insects are then able to deposit pollinia, completing the pollination process. This elaborate procedure virtually ensures cross-pollination and demonstrates one of the many specializations of orchids. Ackerman and Mesler identified frequent pollinators as fungus gnats belonging to the families Mycetophilidae and Sciaridae.

Distribution

Listera cordata is the most widely distributed species of the genus, occurring across much of the Northern Hemisphere. In the United States it is found in the Great Lakes region, the Northeast, the upper Pacific Coast from northern California northward, and the Rocky Mountain region. *Listera cordata* is one of the more common orchids in northern latitudes that range southward along the Rocky Mountain region, and its occurrence in New Mexico is at the southern limit of its range. In New Mexico it occurs in the counties of Rio Arriba, Mora, San Miguel, and Taos. The heart-leaved twayblade has not yet been documented in Arizona.

Habitat

Listera cordata grows in slightly to fairly damp places in pine, fir, and aspen forest at elevations between 9000 and 10,300 feet (2740 and 3150 meters). It grows on flat to gently sloping terrain, usually in moderate to light shade. Often it grows in open forest in a habitat similar to that of *Calypso bulbosa* or *Goodyera repens*. The habitat appears slightly damper because of the relative abundance of herbaceous growth and perhaps the presence of slightly more mosses on the forest floor. Greater concentrations of plants accumulate near the edges of streams, seeps, and boggy areas. In these habitats, *L. cordata* roots in mosses or damp duff.

Blooming Season

The blooming season starts about the middle of June and lasts into the middle of July. The ovaries are swollen even on newly opened flowers, and the seed capsules mature and dehisce in a very short time, usually within 5 weeks of blooming. Seeds are sometimes released from lower capsules while upper flowers are still functional.

Other orchids blooming nearby include *Calypso bulbosa, Corallorhiza maculata,* several *Platanthera* species, and *Coeloglossum viride*. *Goodyera oblongifolia* and *G. repens* grow nearby but do not bloom until later in the season.

Conservation

Listera cordata is known from only four localitites in New Mexico but is relatively common in one of them. Because of its worldwide distribution, it is not threatened when considered across its entire range. Portions of its range in New Mexico are protected within designated wilderness areas. Even though relatively rare in the state, it is not on New Mexico's list of rare plants.

Malaxis Swartz

Prodromus Descriptionem Vegetabilium in Indian Occidentalem: 119. 1788.
Etymology: *Malaxis* is Greek for "softening," referring to the soft leaves in this genus.

The genus *Malaxis* (ma-lax′-is) includes over 300 species distributed around the world in both hemispheres. *Malaxis* is closely related to *Liparis*, a genus that occurs in part of the range of *Malaxis* in the United States and elsewhere, although not in Arizona or New Mexico. *Malaxis* is distinguished from *Liparis* by a shorter column and pollinia with tapered ends. *Malaxis* plants are small with one or rarely two leaves, and a terminal spicate inflorescence with dozens of tiny flowers.

Ten species and varieties of *Malaxis* grow in the United States; 4 occur in Arizona and 3 of those also are found in New Mexico. Several other species of *Malaxis* grow in Mexico just south of Arizona and New Mexico and should be sought in these states. All the *Malaxis* species in the Southwest have a single leaf and can be identified readily by gross visual characteristics, as demonstrated in the key.

Key to the Species of *Malaxis*

1. Flowers purple — *M. porphyrea*
1a. Flowers green
   2. Flowers closely appressed to inflorescence axis — *M. soulei*
   2a. Flowers not closely appressed to inflorescence axis
      3. Inflorescence columnar — *M. abieticola*
      3a. Inflorescence corymbose — *M. corymbosa*

Malaxis abieticola Salazar and Arenas
Lindleyana 16 (3): 149. 2001.

Etymology: The specific epithet refers to the high elevation of the species.

Synonymy:
Microstylis tenuis S. Watson. Proceedings American Academy of Arts and Sciences 26: 152. 1891.
Malaxis tenuis (S. Watson) Ames. Proceedings of the Biological Society of Washington 35: 85. 1922.

Common names: none.

Plate 16

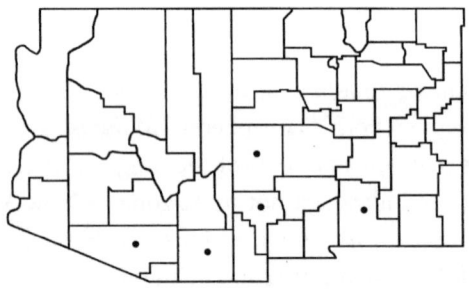

Map 22. Distribution of *Malaxis abieticola*

Description

Plant: to 18 cm tall; single leaf midway on stem, columnar inflorescence with up to 40 tiny, mostly light green flowers.
Roots: a few fibrous roots from cormlike swollen stem.
Leaf: ovate, about 9 cm long and 6.5 cm wide.
Floral bracts: tiny, triangular.
Flowers: green, narrow, 1 cm long × 0.2 cm wide.
Sepals: green; dorsal sepal slightly ovately lanceolate, 5 × 1 mm; lateral sepals narrowly lanceolate, 5 × 0.6 mm.
Petals: green, translucent, filiform, 4 × 0.1 mm.
Lip: green with dark green stripes, arrowhead shape.

Column: 1 mm high, with two pairs of yellow pollinia.
Capsule: narrowly ellipsoidal, about 7 × 2 mm.

Malaxis abieticola (ah-b-a-tea'-cola) grows from a swollen stem base that resembles a corm, with a few fibrous roots. The single ovate leaf is borne about 6 cm above the ground. The inflorescence is cylindrical, from 3 to 5 cm high. A tiny triangular bract is at the base of each petiole. The flowers are delightful but require a powerful hand lens to be fully appreciated. Most are held horizontally with the lip pointing outward, but others face up, down, and sideways. The flowers have a very narrow profile; each one is about 1 cm long but only 2 mm wide. The arrowhead-shaped lip has four dark green stripes, one at each margin and two toward the middle, and is covered with minute papillae. The dorsal sepal is solid green, but the lateral sepals have faint dark green striping and fold together behind the lip, forming an elongated hood. The tiny petals are filiform, translucent, and folded behind the dorsal sepal. The column is less than 1 mm high with two pairs of bright yellow pollinia. The rate of fruit set is very low, perhaps suggesting pollinators are few at the limits of the species' range. Not every plant sets fruit, and those that do have only one to four capsules.

Distribution

Malaxis abieticola grows in much of Mexico, as far south as the state of Oaxaca and the southern tip of Baja California, but in the United States occurs only in New Mexico and Arizona, where it is at its northern limit. Within Arizona it grows only in Cochise and Pima Counties. In New Mexico it is known from only Grant, Otero, and Catron Counties. According to Jennings (2000, personal correspondence), the Grant County record is based on a collection by Greene in 1880 but has not been confirmed since then. The herbarium specimen for Catron County cites Socorro County as the collection location, but it also says it was from the Mogollon Mountains, and that portion of then Socorro County is now part of Catron County.

Habitat

Malaxis abieticola grows in damp mossy and grassy places in fir woods at elevations between 8000 and 9200 feet (2400 and 2800 meters). It also grows near edges of meadows and sometimes out in the meadows, protected from the full sun by taller companion plants of the meadow. It grows on hillsides in the lee of rotting logs and under low-hanging branches of firs.

Blooming Season

Aboveground growth of *M. abieticola* starts in midsummer right after the beginning of the monsoon season, and it blooms about 3 weeks later. If the rains are early, blooming begins in mid-July. Most often, the first flowers open in late July and plants remain in bloom until late August.

Other orchids blooming in the same area include *M. porphyrea*, *M. soulei*, *Platanthera limosa*, *Listera convallarioides*, and *Corallorhiza maculata*, *C. striata*, and *C. wisteriana*. *Malaxis tenuis* and *M. porphyrea* often bloom side by side.

Conservation

Though common in Mexico, *M. abieticola* is rare in Arizona, and even more so in New Mexico. It is known from only two regions in Arizona. In Pima County, several dozen plants grow in a single ravine. In Cochise County, several hundred plants are scattered over a wide region along the crest of the Chiricahua Mountains. Some of its habitat in the Chiricahua Mountains was destroyed by a fierce wildfire in the summer of 1994. *Malaxis abieticola* is ranked G4/S1 in Arizona. The S1 ranking recognizes it as extremely rare in the state. *Malaxis abieticola* now may exist only in one location in New Mexico, as Sivinski and Lightfoot (1995) believe it probably has been extirpated from Otero

County and there are no recent records from Grant County. It is on List 3 but should be moved to List 1. In Arizona, some of its habitat is protected within wilderness areas, but in New Mexico, remaining habitat is threatened by logging.

Notes and Comments

While doing research on the Mexican Orchidaceae, Salazar and Arenas (2001) observed that the combination *Malaxis tenuis* used for one of the small green *Malaxis* in the southwestern United States and Mexico had been previously applied to an Asian species in 1861. They speculate that Oakes Ames was unaware of the earlier use when he transferred *Microstylis tenuis* to *Malaxis* in 1922. Hence a new specific epithet was required for the local taxon and they chose to name it *M. abieticola*.

Malaxis corymbosa (S. Watson) Kuntze
Revisio Generum Plantarum 2: 673. 1891.

Etymology: The name *corymbosa* is derived from the Latin word "corymbus," referring to the broad, flat flower cluster.

Synonymy:
Basionym: *Microstylis corymbosa* S. Watson, Proceedings American Academy of Arts and Sciences 18: 195. 1883.
Achroanthes corymbosa (S. Watson) Greene, Pittonia 2: 184. 1891.

Common names: Madrean adder's mouth.

Plates 17, 18

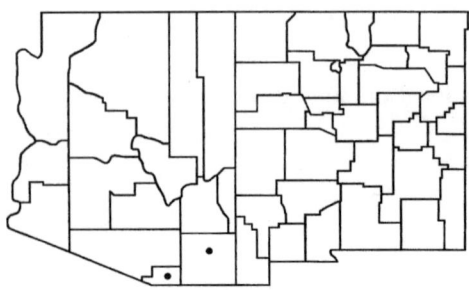

Map 19. Distribution of *Malaxis corymbosa*

Description

Plant: to 26 cm tall, with the top 2 to 3 cm covered with a corymbose arrangement of up to 50 densely packed greenish yellow flowers; a solitary leaf midway on the stem.
Roots: from a round bulblike swollen stem base with a few fibrous roots.
Leaf: cordate to slightly ovate; 6 × 8 cm to 8 × 12 cm.
Floral bracts: pale green, minute, triangular, less than 1 mm long.
Flowers: green, 5.5 × 2.5 mm.
Sepals: green; dorsal sepal nearly linear with a pointed apex, 3 × 1.5 mm; lateral sepals oblong elliptic, 3 × 1.5 mm.
Petals: pale green, filiform, translucent; 0.15 × 2.0 mm.

Lip: rounded at the base without auricles, broadly acuminate in the outer third; 2.5 mm long × 2.0 mm wide.

Column: notched with a pair of bright yellow pollinia on each side of the notch.

Capsule: ellipsoidal, about 0.9 × 0.4 cm.

Malaxis corymbosa (co-rim-bos'-ah) is named for and identified by the shape of its inflorescence. A large tubular bract surrounds the lower third of the stem, and the cordate leaf is midway on the stem. A corymb-shaped raceme of tiny green flowers terminates the stem. The corymb is flat across the top because the flowers are held horizontally, face up. The raceme elongates slightly as it ages, but remains clustered at the top of the stem. The corymbose aspect is maintained until the last flower opens because the pedicels lengthen as the flowers mature. The lips point outward from the center of the raceme, and the flowers are distributed uniformly around the stem, giving the inflorescence a circular appearance when viewed from above. The lateral sepals are borne behind and to the sides of the lip, and the dorsal sepal points directly at the center of the inflorescence. Together the sepals form a broad inverted Y. The lateral petals are so small they are difficult to see without the use of a microscope, and even with a microscope they are elusive because they are often completely hidden under the dorsal sepal. The broad edges of the lip turn up so that it is somewhat cup-shaped and forms a hood around the column. The tip of the lip curves forward slightly.

Distribution

Malaxis corymbosa occurs throughout much of northern Mexico but crosses into the United States only in southeastern Arizona, the northern limit of its range. Within Arizona it grows only in Cochise and Santa Cruz Counties, where it is widely scattered in the Huachuca, Santa Rita, and Chiricahua Mountain ranges. In some areas only a few plants occur, but it is locally common in several canyons. J. G. Lemmon collected the type specimen of *M. corymbosa* in the Huachuca Mountains of southeastern Arizona during July 1882. *Malaxis corymbosa* has not yet been documented from New Mexico.

Habitat

Malaxis corymbosa occurs primarily in mixed oak and pine forest at elevations between 4400 and 7400 feet (1350 and 2250 meters). It grows near streams and springs and in mossy or grassy places in the forest at or near the bottom of a canyon or drainage. Almost all occurrences are within 200 yards (183 meters) of a water source such as a stream, spring, or seep. The streams may be seasonal, but owing to the summer rains they usually have at least a trickle of water when *M. corymbosa* blooms. One particularly beautiful habitat is in mosses among rocks on gently sloping to steep hillsides. In some habitats *M. corymbosa* may be difficult to see because often it is nearly completely hidden under ferns, bushes, or the lower branches of trees.

Blooming Season

Aboveground growth begins shortly after the onset of the monsoon season. The leaves appear as tightly coiled upside-down cones. Even at this early stage of growth, the tight inflorescence is visible in the center of the cone. The leaf uncoils as it lengthens, although it usually still clasps the stem. The heart-shaped leaf eventually flattens out and even droops slightly above the middle. The plants normally bloom within 3 weeks after the summer rains begin, with first bloom as early as the first week in July. The plants continue to open over several weeks, remaining in bloom for about 3 weeks, so the blooming season sometimes extends until the end of August.

Malaxis corymbosa and *M. soulei* often bloom within a few feet of each other, although *M. soulei* typically grows in slightly drier spots. In a few locations, *M. porphyrea* blooms nearby, as does *Platanthera limosa, Schiedeella arizonica,* and *Corallorhiza* species.

Conservation

Since *M. corymbosa* is known from only 2 counties in Arizona, a reasonable conclusion is that it is a rare plant in the United States. There are only about 20 known sites for it in Arizona. However, some sites extend along significant portions of canyons, and colonies contain many hundreds of plants. Because of the relatively large number of individuals, *M. corymbosa* is not considered an endangered species. Part of its range is protected within the U.S. Army's Fort Huachuca installation, and other portions of its range are within designated wilderness areas. *Malaxis corymbosa* is ranked G4/S3S4, with most of its occurrences in Mexico. The state ranking S3 is because just over 20 occurrences are in Arizona.

Malaxis porphyrea (Ridley) Kuntze

Reviso Generum Plantarum 2: 673. 1891.

Etymology: The specific epithet is from the Greek word meaning "purple."

Synonymy:
Microstylis purpurea S. Watson, Proceedings American Academy of Arts and Sciences 18: 195. 1883.
Microstylis porphyrea Ridley, Journal Linnean Society 24: 320. 1888.
Malaxis purpurea (S. Watson) Kuntze. Reviso Generum Plantarum 2: 673. 1891.

Common names: purple malaxis.

Plates 18, 19

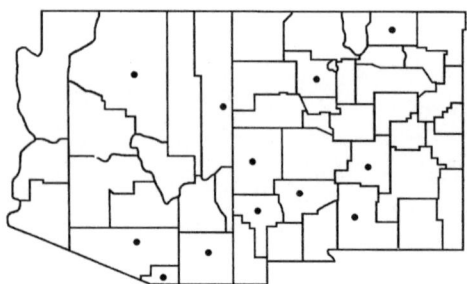

Map 20. Distribution of *Malaxis porphyrea*

Description

Plant: to 34 cm tall with as many as 125 purple flowers; single leaf midway on stem about 5 to 7 cm above ground.
Roots: cormlike structure with a few fibrous roots.
Leaf: ovate, 7 × 4.5 cm.
Flowers: purple, each only 5 × 1.5 mm.
Sepals: purple; dorsal sepal elongated elliptic, 2.5 × 0.5 mm; lateral sepals lanceolate elliptic, 3.0 × 0.6 mm.
Petals: purple, filiform, 3 × 0.1 mm.
Lip: purple with cream to yellowish center, arrowhead-shaped, 3.0 × 2.0 mm.

Column: less than 1 mm high with two pairs of bright yellow pollinia. Capsule: ellipsoidal to almost spherical.

Malaxis porphyrea (por-phy-ree′-ah) grows from a swollen rhizome that develops into a cormlike structure with a few fibrous roots. The flower stem has a ridged, angular cross section with purple tinges lining the ridges. There is a single ovate, acute leaf partway up the stem. The purple flowers are the smallest orchid flowers in Arizona and New Mexico. The flowers angle away from the inflorescence axis, with the dorsal sepal outermost and the lip closest to the stem. The sepals, petals, and lip are covered with tiny papillae that look like raised bumps, or half-spheres. The papillae are not discernable to the unaided eye and are barely visible through a 10-power hand lens. The lip is shaped like a narrow arrowhead, with the margins folded up to meet the column. A raised triangular ridge in the center of the lip rises sharply at the column. The ridge is cream to whitish yellow, and the rest of the lip is purple. The margins of the dorsal sepal fold backward and form a reverse trough. The lateral sepals arch backward slightly. The translucent petals are delicate and fold backward to cross behind the ovary. They are so tiny they are all but invisible to the unaided eye. The capsules are ellipsoidal to almost spherical. The rate of fruit set is very low; few plants fruit, and those that do produce only one or two capsules per plant.

Distribution

Malaxis porphyrea grows in northern Mexico and in Arizona and New Mexico. Within Arizona it grows in the 5 counties of Apache, Cochise, Coconino, Pima, and Santa Cruz. It apparently skips the counties of Graham and Greenlee, which lie between Cochise and Apache Counties. Most likely it is in both of those counties but has not yet been documented there. It is at its northern limit in New Mexico where it is in Catron, Colfax, Grant, Lincoln, Otero, Sandoval, and Sierra Counties. William Jennings (2000, personal correspondence) additionally reported it from San Miguel and Los Alamos Counties. The type specimen of *M. porphyrea* was collected in the Huachuca Mountains of Arizona.

Habitat

Malaxis porphyrea grows in mixed oak, fir, and pine forest at elevations between 7000 and 9200 feet (2100 and 2800 meters). Its most common habitat is near damp, mossy or grassy places in slightly open areas in the forest. These locations can be recognized because they have more herbaceous growth than surrounding parts of the forest. It also grows near and in the edges of meadows and sometimes in open meadows, sheltered from full sun by taller grasses.

Blooming Season

Aboveground growth of *M. porphyrea* starts about the beginning of July with the onset of the monsoon rains. The earliest recorded bloom is 14 July, most likely in a year when the monsoons started early. Usually first blooming occurs in late July, and the plants stay in bloom through August and into early September in years when the rains arrive late.

Malaxis porphyrea blooms near *M. abieticola*, *M. corymbosa*, *M. soulei*, *Goodyera oblongifolia*, and *Platanthera limosa*. Earlier in the year *Schiedeella arizonica* blooms nearby, as do *Corallorhiza maculata*, *C. striata*, and *C. wisteriana*.

Conservation

Malaxis porphyrea is widely scattered in Arizona and New Mexico and exists in large colonies in a few locations. However, it is rare and needs protection in both states. Campers digging in a meadow destroyed a major colony in the Chiricahua Mountains of Cochise County, Arizona, during 1999. In Arizona it is ranked G2G3/S2 because there are fewer than 20 occurrences in the state. It is not listed in New Mexico but should be on List 1 because it is so rare. Part of its range in Arizona's Huachuca Mountains is contained within the Fort Huachuca Army Base. Other portions of its range in both states are designated wilderness areas and therefore are safe from development, logging, and mining.

Notes and Comments

The plants in the United States and parts of northern Mexico that are known now as *M. porphyrea* were known for many years as *M. ehrenbergii* (Reichenback f.) Kuntze. J. G. and S. P. Lemmon originally discovered the species in 1882 in the Huachuca Mountains, and S. Watson named it *Microstylis purpurea*. Ridley recognized that the name *Microstylis purpurea* had been previously used for another plant, and in 1888 he named the taxon *Microstylis porphyrea*. Kuntze later transferred it to *Malaxis*. Ames (1924) placed it in synonymy with *M. ehrenbergii*. Most references to the purple malaxis in the United States after 1924 followed Ames, and used the name *M. ehrenbergii* until Salazar (1993) described a new species he called *M. wendtii*. Salazar separated the species based on lip shape and the presence of papillae. The lip of *M. wendtii* has the shape of a long narrow acuminate arrowhead, while that of *M. ehrenbergii* is broadly triangular hastate. The floral parts of *M. wendtii* are covered with papillae, and *M. ehrenbergii* does not have them. Salazar said that the plant had not been described previously and grew only in the Mexican states of Coahuila and Nuevo Leon. Todsen (1995) then pointed out that all of the plants in the United States formerly considered to be *M. ehrenbergii* were actually *M. wendtii*. In a subsequent paper Todsen (1997) revised his position, recognizing that *M. wendtii* differed from the more northern plants in Mexico and those in the United States. Since it was no longer considered synonymous with either *M. ehrenbergii* or *M. wendtii*, Kuntze's revision of Watson's original description has precedence, and Todsen stated that the taxon in Arizona and New Mexico is correctly known as *M. porphyrea*. *Malaxis wendtii* occurs in the United States only in the Chisos Mountains of Big Bend National Park, Brewster County, Texas.

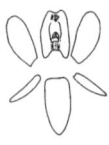

Malaxis soulei L. O. Williams
Annals Missouri Botanical Garden 21: 343. 1934.

Etymology: The species was named for Justus F. Soule, a professor at the University of Wyoming.

Synonymy:
Microstylis montana Rothrock. Report upon United States Geographical Surveys West of the One Hundredth Meridian, Vol. VI — Botany: 264. 1878.
Malaxis montana (Rothrock) Kuntze. Reviso Generum Plantarum 2: 673. 1891.
Achroanthes montana Greene. Pittonia 2: 183. 1891.

Common names: mountain malaxis, rat-tailed malaxis.

Plate 19

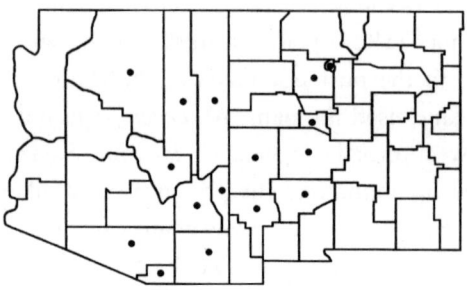

Map 21. Distribution of *Malaxis soulei*

Description

Plant: up to 35 cm tall, single leaf, spicate inflorescence, with tiny flowers tightly appressed against inflorescence axis.
Roots: a few fibrous roots from bulblike base.
Leaf: ovate to elliptic to linear elliptic, up to 12 × 4.5 cm.
Flowers: green, to yellowish green; 4.5 mm long × 2.5 mm wide.
Sepals: green to greenish yellow; dorsal sepal linear and rounded gently at the apex, about 1.2 × 2.5 mm; lateral sepals oblong ovate, about 1.0 × 2.0 mm.

Petals: green, filiform, 0.2 × 2.0 mm, usually curved around behind the dorsal sepal.

Lip: green to yellowish; 1 mm wide × 2 mm long; three-lobed, notched at top and bottom.

Column: less than 1 mm high; two pairs of pollinia, yellow, translucent.

Capsule: spheroidal to elliptic.

Malaxis soulei (soo lee′-eye) grows from a round cormlike swollen stem base with a few fibrous roots. A new corm forms while the old one is still present, so during part of the year, there is a pair of corms. The lower portion of the stem is totally sheathed by a tubelike bract. The solitary leaf starts about 4 cm above the ground. The leaf varies from ovate to elliptic to linear elliptic and is sometimes rounded at the apex and sometimes slightly pointed. The spicate inflorescence is densely covered with more than 100 tiny flowers. The structure of the inflorescence varies from plant to plant. On some the flowers start within 2 cm of the base of the leaf. On others the first flowers are 7 cm above the leaf. Up to 20 cm of the raceme bears flowers, but it is only 6 mm thick, earning it the common name of rat-tail malaxis. Flower color varies from light yellowish green to dark green, to bicolored, marked with dark green on the inside of the petals and on the dorsal sepal and light green elsewhere. Some plants have both light and dark green flowers, with the light green ones on the bottom half of the inflorescence and the dark ones on the upper half. The differently colored flowers start to bloom at the same time, with the light green ones blooming from the bottom up and the dark green ones starting to bloom in the middle of the stem and continuing to the top.

The flowers are not resupinate so the three-lobed lip is uppermost on the flower. The side lobes of the lip are spreading and inrolled, creating an open trough on freshly opened flowers. After pollination the side lobes fold together, blocking access to the column. The lip is deeply notched at the upper end, with a minute tooth in the notch and lower ears on the lateral lobes that extend well below the junction with the column. The dorsal sepal is linear and rounded gently at the apex. The lateral sepals are oblong ovate and spread behind the lip. The filiform petals are so tiny they are difficult to see with the unaided eye, and are usually curled around behind the dorsal sepal.

Distribution

Malaxis soulei grows in northern Mexico, Texas, New Mexico, and Arizona. In Arizona it is in the 9 counties of Apache, Cochise, Coconino, Gila, Graham, Greenlee, Navajo, Pima, and Santa Cruz. The type specimen of *M. soulei* was collected from Mt. Graham in Graham County, Arizona, by J. T. Rothrock in 1874, during a botanical expedition led by Lt. Wheeler. In New Mexico, it is found in the 7 counties of Catron, Colfax, Grant, Los Alamos, Sandoval, Sierra, and Valencia. William Jennings (2000, personal correspondence) reported it also in Colfax and San Miguel Counties. It is at its northern limit in New Mexico.

Habitat

Malaxis soulei grows in mixed or pure stands of oak, juniper, pine, and fir between the elevations of 5300 and 9200 feet (1600 and 2800 meters). It adapts to a wide range of soil moisture conditions, from open, dry sites to grassy and mossy places in the forest. It often grows near the edges of meadows, on steep hillsides, and on open, rocky slopes. At the lower elevations of its range, it grows in dense shade under low-hanging branches of junipers.

Blooming Season

The first leaves of *M. soulei* appear shortly after the start of the monsoon season. The spikes elongate rapidly, and the first flowers open about 3 weeks later, as soon as mid-July if the rains started early. Additional plants continue to emerge throughout the summer, resulting in a long flowering season. Blooming plants persist until the first few weeks of September, although others in the area will already have mature fruit. Sometimes dense colonies of blooming plants form, with dozens in an area the size of a large dinner plate. Fruit set varies widely. Some plants set no fruit while others set only 1 to 10 capsules; occasionally a stem will be totally covered with fruit.

Because of its wide tolerance for elevation and habitats, *M. soulei* blooms near several other orchids. At higher elevations, *M. soulei*, *M. abieticola*, and *M. porphyrea* often bloom within a few yards of each other and at the same time. It also grows in the same habitats as *Calypso bulbosa*, *Goodyera oblongifolia*, *Schiedeella arizonica*, *Corallorhiza maculata*, *C. striata*, and *C. wisteriana*, although some bloom at different times. At middle elevations, it blooms at the same time and near *M. corymbosa*. At the lowest end of its elevation range, *M. soulei* is the only other orchid associated with *Stenorrhynchos michuacanum*.

Conservation

Malaxis soulei is the most widespread and numerous of the genus in Arizona and New Mexico. Its affinity for different habitats and high elevational tolerance mean it is safe from threats at this time. Much of its habitat is protected within designated wilderness areas.

Notes and Comments

J. T. Rothrock placed the plant he discovered in the genus *Microstylis*, which was later determined to be a synonym for *Malaxis*. However, the specific epithet *montana* had been used nearly 50 years earlier for a species from Java, and therefore could not be applied to Rothrock's specimen. L. O. Williams (1934), perhaps recognizing the precedence of the Javan *M. montana*, renamed our plant *M. soulei*. However, he (1937) soon placed the name *M. soulei* in synonymy with *M. macrostachya* (Lexarza) Kuntze. Still later, Williams (1951) went back to the use of *M. soulei*, noting that the type specimen for *M. macrostachya* was not in existence, and the written descriptions did not identify it with certainty. Luer (1975) used *M. macrostachya* for our plant, but more recent treatments of *Malaxis* (McVaugh 1985, Soto 1988) use *M. soulei*, relegating *M. macrostachya* to a synonym for *M. carnosa*.

Piperia Rydberg

Bulletin Torrey Botanical Club 28: 269. 1901.
Etymology: The genus was named in honor of Professor C. V. Piper of the Agricultural Experiment Station at Pullman, Washington.

Piperia (pye-per'-ee-ah), with 10 species, is a relatively large genus by North American standards. The center of distribution for the genus is in California, and several of the species are endemic there. Taxonomic treatments of *Piperia* have varied widely over the years. Rydberg (1901) separated *Piperia* from the genus *Habenaria* on the basis of morphology and pollination mechanism, recognizing 9 species. Ames (1910) included the plants within *Habenaria* and recognized only *P. elegans* and *P. unalascensis*. Correll (1950) also included the plants within *Habenaria*, taking the extreme position of recognizing only 1 species and lumping all taxa as variants of *P. unalascensis*. Luer (1975) returned to the concepts proposed by Rydberg, separating *Piperia* from *Habenaria* and recognizing 4 species. Ackerman (1977) revised the genus, identifying 4 species and 1 subspecies. Thirteen years later, Morgan and Ackerman (1990) showed that the short-spurred plants usually lumped together under *P. unalascensis* consisted of at least 4 sibling species, 2 of which they described for the first time. Additionally, they raised the status of the subspecies Ackerman had identified in 1977 to a full species. Morgan and Glicenstein (1993) added 1 more species and a subspecies. Only *P. unalascensis* is known from this area, and only in New Mexico.

One primary distinguishing characteristic of *Piperia* plants is that the leaves are frequently faded at anthesis. Their annual cycle starts in the fall with the formation of new roots, one of which will produce a replacement tuber. The two to six leaves usually appear in early spring, and in most species the flower spikes develop in late spring and early summer. As each spike matures, the leaves turn yellow and wither away. Often only a twisted brown trace of the leaves endures as the buds open. Each plant remains in flower for 4 to 6 weeks.

Piperia unalascensis (Sprengel) Rydberg
Bulletin Torrey Botanical Club 28: 270. 1901.

Etymology: This species is named for the Aleutian Island Unalaska where it was first discovered.

Synonymy:
Basionym: *Spiranthes unalascensis* Sprengel, Systema Vegetabilium 3: 708. 1826.
Habenaria schischmareffiana Chamisso, Linnaea 3: 29. 1828.
Platanthera foetida Geyer ex Hooker, Kew Journal of Botany 7: 376. 1855.
Herminium unalascensis (Sprengel) Reichenbach, Orchids of Europe: 107. 1851.
Habenaria unalaschcensis (Sprengel) S. Watson, Proceedings American Academy of Arts and Sciences 12: 277. 1877.
Platanthera unalaschcensis (Sprengel) Kurtz, Botanische Jahrbücher für Systematik 19: 408. 1894.
Montolivaea unalaschcensis (Sprengel) Rydberg, Memoirs New York Botanical Garden 1: 107. 1900.

Common names: Alaska piperia, slender-spire orchid.

Plate 20

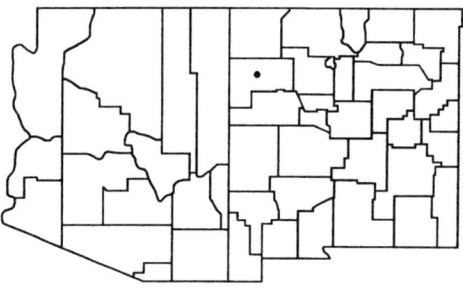

Map 23. Distribution of *Piperia unalascensis*

Description

Plant: between 15 and 70 cm tall elsewhere, but about 15 to 40 cm tall in this region; two to six basal leaves withering at anthesis, up to 100 tiny flowers.

Roots: tuberoid; usually in pairs consisting of current and former year's growth.

Leaves: oblanceolate to elliptic, 8 to 10 cm × 2 to 3 cm.

Floral bracts: minute, ovate.

Flowers: green, musty aroma, about 0.5 × 0.6 cm.

Sepals: green, ovate lanceolate to ovate elliptic, 3 × 1.5 mm.

Petals: green, ovate lanceolate, 3 × 1 mm.

Lip: green, ovate lanceolate to triangular ovate, curls upward at tip, 4 × 2.5 mm.

Spur: cylindrical, about as long as lip.

Column: minute.

Capsule: ellipsoidal.

Piperia unalascensis (un-a-las-sen'-sis) is fairly easy to identify because of plant structure. The spurred green flowers bear some resemblance to *Platanthera* flowers, and the loosely flowered inflorescence resembles that of *Platanthera purpurascens* from a distance. Both plants grow in mesic conifer forests. However, the leaves of *P. purpurascens* are proportionally larger, more linear, scattered more along the stem, and fresh and green at blooming. The leaves on *Piperia unalascensis* are primarily basal and are faded or fading at anthesis. The flowers are relatively small. The dorsal sepal and petals are free and spreading, a feature that helps distinguish the flowers from those of *Platanthera*, where the dorsal sepal and petals form a hood over the column. The margins of the lip near the entrance to the nectary are turned up slightly to form a guide for the pollinator. The lateral sepals are reflexed and clasp the spur. The spur varies in length from slightly shorter than the lip to slightly longer than the lip, varying in position from horizontal to decurrent. The flowers have a faint musty aroma, which some find unpleasant. The ellipsoidal capsules are held upright along the stem.

Ackerman (1977) reported that *Piperia unalascensis* is pollinated by both pyralid moths (*Oidaematophorus* species) and plume moths (*Platyptilia* species). Ackerman said the flowers are "protandrous by age-dependent lip movement." On newly opened flowers, the lip is tightly held to the column and covers the entrance to the nectary. Visiting moths trying to probe for nectar contact the viscidium and remove pollen. As the flower ages, the lip moves downward, exposing the entrance to the nectary and allowing pollen deposition. However, by this stage, lip movement has effectively moved the viscidium out of reach, and the anther sacs have dried and shriveled, making pollen removal nearly impossible.

Distribution

Piperia unalascensis is the most far-ranging of the genus. It grows from the mountains of southern California to the Aleutian Islands, and eastward to Quebec, Canada. It also grows in the northern Lake Huron region (Whiting and Catling 1986) and in much of the Rocky Mountains. *Piperia unalascensis* does not occur in Arizona and is known in New Mexico only from a single wooded canyon bottom in McKinley County where it was collected in 1987. This is the southeastern limit of its range.

Habitat

Piperia unalascensis grows at about 8000 feet (2500 meters) in New Mexico. In other portions of its range, it grows near sea level in California and at 10,000 feet (3050 meters) in Utah (Szczawinski 1975). It grows in mesic conditions in open mixed or coniferous forest in bright to moderately shaded conditions. In some areas the plants grow in full sun. This species should be sought in mountainous regions in northern Arizona and northern New Mexico because the habitats in many areas in both states are ideal.

Blooming Season

Piperia unalascensis is in bloom between early May and the end of August in most of its range. Its blooming season in New Mexico is conservatively estimated as the month of July based on the single observation. If more sites are located, the blooming season in New Mexico can be established with greater accuracy. *Corallorhiza maculata* blooms not too far from the reported location of *P. unalascensis*, but in Cibola County. No other orchids are known to grow in McKinley County.

Conservation

Piperia unalascensis is locally common and widely scattered in many parts of its total range. However, it is extremely rare at this limit of its range in New Mexico. The author was unable to locate it in 1999, so it may have been extirpated from New Mexico or is just extremely well hidden. It is on List 3 of New Mexico's list of rare plants, implying more data are needed, but it belongs on List 1.

Platanthera L. C. Richard

Mémoires du Museum d'Histoire Naturelle Paris 4: 48. 1818.
Etymology: *Platanthera* derives from Greek words meaning "wide anther," in reference to the broad anther on the flowers in this genus.

Platanthera (pla-tan'-ther-ah), a confusing genus still in revision, was established by Richard in 1818, but his proposal was not universally followed. Orchid researchers for the first half of the twentieth century such as Ames (1924) and Correll (1950) did not recognize *Platanthera*. However, in his work on the orchids of North America, Luer (1975) recognized *Platanthera* as one of three orchid genera often included in the once-cumbersome *Habenaria*. The other two are *Piperia* and *Coeloglossum*. *Platanthera* species can be readily distinguished from *Coeloglossum* by the acute vice trilobed lip, and from *Piperia* species by leaves that remain green long past flowering. *Platanthera*, *Coeloglossum*, and *Piperia* are represented in the combined orchid flora of Arizona and New Mexico. In the United States *Habenaria* is found only in the extreme Southeast.

Seven *Platanthera* taxa occur in Arizona and New Mexico, more than from any other orchid genus in the two states. All of these species are in what is often referred to as the *dilatata-hyperborea* complex or the *Limnorchis* section of *Platanthera*. The section name is based on the genus *Limnorchis* that Rydberg (1901) created for these and several related species. This complex has experienced a wide variety of treatments. Rydberg's approach represented one extreme. At the other extreme, Ames (1924) reduced many members of the *dilatata-hyperborea* complex to varieties of *P. hyperborea*. The taxonomy is still being debated, and Weber (1989) advocated the use of Rydberg's *Limnorchis*.

Platanthera dilatata is excluded from this region based on extensive herbarium and field research leading to this volume. One herbarium record suggests *P. dilatata* var. *leucostachys* may be in northern Arizona, but the exact location of the collection is unclear and may have been in Utah, where *P. dilatata* var. *leucostachys* is known. Three herbarium records from New Mexico tentatively identified as *P. dilatata* are all traceable to *P. huronensis*. However, due to its proximity in Colorado, *P. dilatata* var. *albiflora* someday may be positively located in New Mexico.

The key below attempts to use characters familiar to laymen and visible to the unaided eye or with just a hand lens. The extremes of all the listed species can be identified by use of this key. However, intermediate forms of some species such as *P. purpurascens*, *P. aquilonis*, and *P. sparsiflora* are apt to confuse even the expert, a possible indication of hybridization. Sheviak (1999b) suggested another reason for the extreme variation within species is regional specialization to pollinators or other local factors. Dried herbarium specimens of some of these species are especially difficult to differentiate. J. T. Atwood (1987, personal communication) summed it up concisely when he observed, "These plants do not seem to fit species concepts very well." Table 9 at the end of this section summarizes key features of each species.

Key to the Species of *Platanthera*

1. Leaves reduced to bracts along stem — *P. brevifolia*
1a. Leaves well formed
    2. Column large, half or more as large as hood formed by dorsal sepal and petals
        3. Lip linear to linear lanceolate; spur slightly shorter than lip to 1.5 times lip length — *P. sparsiflora*
        3a. Lip linear to linear elliptic; spur greater than 1.5 times lip length — *P. zothecina*
    2a. Column small, less than half as large as hood formed by dorsal sepal and petals

4. Flowers self-pollinating, with pollinia rotated forward and spilled onto stigma
    5. Lip rhombic to rhombic lanceolate, yellowish — *P. aquilonis*
    5a. Lip dilated at base, acuminate, whitish — *P. huronensis*
4a. Flowers not self-pollinating, and pollinia within anther sacs unless scattered by pollinator
    6. Flowers distinctly whitish green — *P. huronensis*
    6a. Flowers green, to yellow-green, to bluish green
        7. Spur at least 1.5 times as long as lip — *P. limosa*
        7a. Spur much less than 1.5 times as long as lip — *P. purpurascens*

Platanthera aquilonis Sheviak
Lindleyana 14 (4): 193–194. 1999.

Etymology: The name *aquilonis* means "of the north," in reference to its range and also to maintain identity with *P. hyperborea*, the name previously commonly used for this species.

Synonymy: none.

Common name: green bog-orchid.

Plate 21

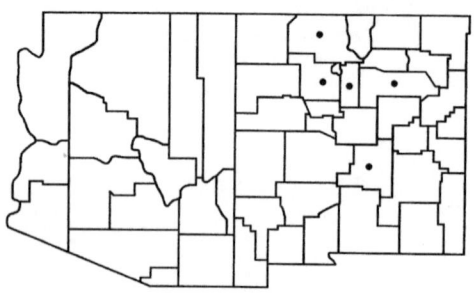

Map 24. Distribution of *Platanthera aquilonis*

Description

Plant: 30 to 40 cm tall, leaves alternating, ascending along stem, gradually reduced to bracts above; raceme with up to 50 flowers; plants larger and with more flowers in other portions of range.

Roots: few, thick, short, 2 to 5 cm long.

Leaves: four or more, oblong to linear lanceolate, to 10 cm long × 2 cm wide, replaced higher on stem with lanceolate bracts.

Floral bracts: lanceolate, to 1.5 cm long on lower flowers, shorter above.

Flowers: slightly yellowish green; self-pollinating.

Sepals: dorsal sepal green, ovate, 3 mm wide × 3.5 mm long; lateral sepals green, slightly falcate, broadly lanceolate, folded down along spur, 4.2 × 2.0 mm.

Petals: yellow-green, narrowly lanceolate, with basal half broadly triangular tapering sharply to narrow apex.

Lip: yellow-green, rhombic lanceolate to lanceolate but not conspicuously dilated, not acuminate, often held nearly perpendicular to column, 4.5 × 1.5 mm.

Column: wider than high, less than half as large as hood formed by dorsal sepal and petals, anther sacs diverging toward bottom; pollinia spilled onto stigma even on unopened flowers.

Spur: yellow-green, shorter than lip, curved, clavate to cylindrical with rounded tip, 3.5 mm long × 1 mm thick.

Capsule: ellipsoidal.

Platanthera aquilonis (ak-wi-lon'-iss) was considered by many authors (Ames 1924, Correll 1950, Luer 1975) as highly variable (although they discussed it as either *Habenaria hyperborea* or *P. hyperborea*), and across its entire range, it does show more variation than is evident in the southwestern states. Here, it is rather uniform in appearance, and the yellow-green flowers can be separated from related species such as *P. purpurascens* by its rhombic lanceolate lip, lack of aroma, and completely self-pollinating habit. The hood formed by the dorsal sepal and petals is large compared to the column, and the petals are usually curled back along the spur, which is shorter than the lip, clavate, curved, and blunt. The column is perhaps the best single feature to use to identify the species. It can be studied with the unaided eye but is more easily observed with a 10-power hand lens. The column is wider than it is high, and the anther sacs diverge widely after nearly touching at their apices. Even on flowers that have not yet opened, the anther sacs are split, and the pollinia have rotated down onto the stigma.

Platanthera huronensis is perhaps the most closely related species but can be distinguished by flower presentation and color. On *P. aquilonis*, the hood formed by the dorsal sepal and petals is nearly horizontal, and almost perpendicular to the lip. *Platanthera huronensis* displays the flowers vertically, with the hood aligned with the lip. *Platanthera aquilonis* is green, although the lip and spur are yellowish green. *Platanthera huronensis* is a very whitish green, especially in the lip and spur. Albino forms of *P. aquilonis*, with pure white plant and flowers such as reported from Canada (Light and MacConaill 1989, Szczawinski 1975) might make distinguishing between *P. aquilonis* and *P. huronensis* difficult based on flower color, but the white form of *P. aquilonis* is not known to occur in the Southwest.

Distribution

Platanthera aquilonis is by far the most widely distributed member of the genus. Sheviak (1999b) reported it as transcontinental in the North, even extending beyond the Arctic Circle. In the United States it occurs in the Northeast, the Great Lakes region, the Rocky Mountain states, and along the northern Pacific Coast. It is rather infrequent in New Mexico, occurring only in the counties of Lincoln, Rio Arriba, Sandoval, Santa Fe, and San Miguel, and does not occur in Arizona. It is at its southern limit in New Mexico. Published reports of *P. aquilonis* (as *Habenaria hyperborea*) in Arizona, such as those in Kearney and Peebles (1951) and Lehr (1978), are referable to *P. purpurascens*.

Habitat

Platanthera aquilonis grows in wet areas at elevations between 8000 and 11,200 feet (2340 and 3400 meters). In high mountain meadows and roadside ditches, it grows in full sun. Other typical locations are on wet banks of streams and hillside seeps in bright to direct light. Still others are marshy, swale-type areas bordering lakes and rivers. Because of its small stature in the Southwest, it may be well hidden by taller grasses in all of these habitats.

Blooming Season

Depending on elevation, *P. aquilonis* can be found in bloom from mid-June to the first part of August, and can be counted on to be in bloom in July. In all of its habitats it may grow with other members of the genus and *Spiranthes romanzoffiana*.

Conservation

Though infrequent in New Mexico, *P. aquilonis* grows in diverse habitats and is sufficiently scattered to be safe from threats. Its flowers are not particularly showy and blend in well with its lush environments, and so are hard to discover. Hence, they are relatively safe from that bane of many of the more showy orchids: those who would dig them up for transplanting to home gardens.

Notes and Comments

While doing research on *Platanthera* for the Flora of North America project, Sheviak (1999b) concluded that the North American plants that had long been identified as either *Habenaria hyperborea* var. *hyperborea* or *Platanthera hyperborea* var. *hyperborea* were a totally different and unnamed species. The specific epithet *hyperborea* had been applied erroneously to these plants for over 100 years. Originally described by Linnaeus as *Orchis hyperborea* based on material from Iceland, *P. hyperborea* differs from the North American plants in chromosome number, column structure, and lip shape. Sheviak also determined that none of the numerous synonyms associated with *P. hyperborea* could be traced to the North American plants. He chose the new name *P. aquilonis* to maintain the association with the North based on the old name *P. hyperborea*, which roughly means "above the North."

Platanthera brevifolia (Greene) Kränzlin
Orchidacearum Genera et Species I: 639. 1899.

Etymology: The specific epithet refers to the shortened leaves of this species.

Synonymy:
Basionym: *Habenaria brevifolia* Greene, Botanical Gazette 6: 218. 1881.
Limnorchis brevifolia (Greene) Rydberg, Bulletin Torrey Botanical Club 28: 631. 1901.
Habenaria sparsiflora S. Watson var. *brevifolia* (Greene) Correll, Leaflets of Western Botany 3: 244. 1943.
Platanthera sparsiflora (S. Watson) Schlechter var. *brevifolia* (Greene) Luer, Native Orchids of the United States and Canada: 240. 1975.

Common names: none.

Plate 22

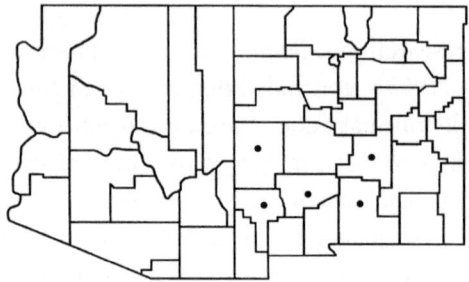

Map 25. Distribution of *Platanthera brevifolia*

Description

Plant: 20 to 40 cm tall with up to 40 flowers; appears leafless.
Roots: few, fibrous, from thickened rhizome.
Leaves: four or more, alternating, scattered along stem, reduced to mere bracts, sheathing stem; each about 3.5 × 1 cm.
Floral bracts: lanceolate, at each flower, about 2.5 cm long × 0.6 cm wide.
Flowers: green, about 2 cm from tip of dorsal sepal to end of spur, about 1.5 cm wide; dorsal sepal and petals form hood over column.

Sepals: dorsal sepal ovate, 4 × 4.5 mm; lateral sepals elliptic lanceolate, slightly oblique, up to 0.3 × 1.0 cm.

Petals: falcate, 3 mm wide × 6 mm long.

Lip: lanceolate to slightly elliptic lanceolate, 0.3 × 1.0 mm, with barely raised central ridge running length of lip; apex upturned.

Spur: cylindrical, acute, slightly curved, as long as 1.5 times the lip, to 1.2 mm thick × 1.5 mm long.

Column: 4 × 4 mm, anther sacs nearly parallel, with yellow, granular pollen.

Capsule: ellipsoidal.

Platanthera brevifolia (brev-i-fo´-lee-ah) is easily identified by the leaves, which are reduced to bracts clasping the stem. More than half of the stem is taken up by the loosely to densely flowered inflorescence. The flowers are green to greenish yellow. The long cylindrical spur dwarfs the lip. As with all the *Platanthera* in the *Limnorchis* section, the dorsal sepal and petals form a hood over the column. The large, broad column occupies over half the hood formed by the dorsal sepal and petals, a characteristic in common with *P. sparsiflora* and *P. zothecina*. The sepals are widely spreading on newly opened flowers but may reflex as they mature. The lip is long and linear to linear elliptic and has a slightly raised ridge running lengthwise down the center. *Platanthera sparsiflora* has a short raised knee running vertically down the lip near the opening to the spur. The differences in the lip are more apparent in fresh material than in dried herbarium specimens.

The flowers are disproportionally large for the size of the stem, one of the characteristics that set *P. brevifolia* apart from *P. sparsiflora* and *P. zothecina*. While flower size overlaps in these 3 related species, the largest examples of *P. brevifolia* exceed the largest of either *P. sparsiflora* or *P. zothecina*.

Distribution

Platanthera brevifolia is widespread in Mexico, but in the United States occurs only in 5 counties of New Mexico: Catron, Grant, Lincoln,

Otero, and Sierra. The location of the type specimen is in Grant County. Its relative proximity to the Arizona border, and the seemingly suitable habitat in the Chiricahua and Pinaleño Mountains of southeastern Arizona, suggest *P. brevifolia* should also grow there. However, there are no records of it from the state, and searches conducted as part of the fieldwork for this book failed to find any plants in Arizona. Within New Mexico, *P. brevifolia* is at its northern limit and replaces *P. sparsiflora* in the southern part of the state.

Habitat

Platanthera brevifolia grows in pine-oak woodland and in pure conifer forest at elevations between 6500 and 9000 feet (1980 and 2700 meters). It is found above 10,000 feet (3050 meters) in Mexico. The typical habitat is partial shade in open forest within 50 to 100 yards (45 to 90 meters) of a seasonal drainage or minor stream. It does not require the constantly moist environment of most western *Platanthera* species, preferring dry sites, even when close to a stream. Most often it is on flat to slightly sloped areas, but it also grows on steep slopes. It does not form large colonies as do some of the other *Platanthera* species, appearing as scattered individuals or in small groups of 10 to 12 plants.

Blooming Season

The blooming season stretches from 1 July to 25 September, but flowering in September is unusual and in most years the season is limited to July and August. Other orchids growing near *P. brevifolia* may include *Schiedeella arizonica*, *Malaxis porphyrea*, *M. soulei*, *Corallorhiza maculata*, and *C. striata*. All but the 2 *Malaxis* species are usually past flowering by the time *P. brevifolia* opens.

Conservation

Platanthera brevifolia is widespread and locally common in New Mexico, and more plentiful in Mexico. Its distribution and habitat suggest it is safe from threats at this time.

Notes and Comments

Treatment of *P. brevifolia* has varied over the years. After it was described in 1881, early authors such as Ames (1910, 1924) maintained its specific status. Starting in the middle of the twentieth century, first Correll (1943) and then Luer (1975) and McVaugh (1985) treated it as a variety of *P. sparsiflora*. Both Ames and Correll considered it a *Habenaria*, consistent with the treatment prevalent at the time, while Luer and McVaugh placed it in *Platanthera*. It is treated here as a separate species based on habitat, plant structure, size of the flower, and details of the lip.

Platanthera huronensis (Nuttall) Lindley
Genera and Species of Orchidaceous Plants: 288. 1835.

Etymology: Named after Lake Huron, near where it was first discovered.

Synonymy:
Basionym: *Orchis huronensis* Nuttall, Genera of North American Plants 2: 189. 1818.
Habenaria huronensis (Nuttall) Sprengel, Systema Vegetabilium 3: 688. 1826.
Limnorchis huronensis (Nuttall) Rydberg, in Britton, Manual of the Flora of the Northern States: 294. 1901.
Habenaria hyperborea var. *huronensis* (Nuttall) Farwell, Papers Michigan Academy of Science I: 92. 1923.
Platanthera hyperborea var. *huronensis* (Nuttall) Luer, Native Orchids of the United States and Canada: 230. 1975.
Platanthera × *media* (Rydberg) Luer, Native Orchids of the United States and Canada: 229. 1975.

Common names: tall northern green orchid, tall green bog-orchid.

Plates 22, 23

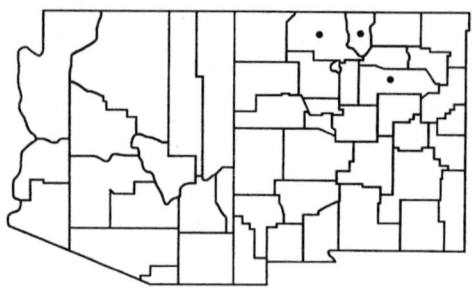

Map 26. Distribution of *Platanthera huronensis*

Description

Plant: up to 100 cm tall in parts of its range, more typically 20 to 70 cm tall here; leaves alternating, ascending along stem, gradually reduced to bracts higher on stem; raceme densely flowered with 50 to 100 flowers.

Roots: few, thick, fibrous, from thickened rhizome.

Leaves: five or six, oblong to linear lanceolate, basal leaf rounded at tip, others pointed; to 20 × 3 cm.

Floral bracts: lanceolate, largest 2 × 0.3 cm, smaller higher on stem.

Flowers: whitish green, each about 1 to 1.5 cm × 0.7 to 1.5 cm.

Sepals: greenish white; dorsal sepal ovate, acute, slightly concave, 5 × 3 mm; lateral sepals spreading, falcate, 5 × 1.8 mm.

Petals: whitish green with three faint green veins along length, falcate, 5 × 1.5 mm.

Lip: whitish green; mostly lanceolate with slightly dilated base; 8 mm long × 2 mm wide.

Spur: slightly curved, about as long as lip, cylindrical, with rounded tip, 6 mm long × 0.9 mm thick.

Column: longer than wide, 5 × 3 mm; anther sacs nearly parallel; pollen granular.

Capsule: ellipsoidal, erect, 1 cm long and 0.5 cm in diameter.

Platanthera huronensis (hur-on-en'-sis) can be identified on sight by its color. It is distinctly whitish green and stands out from the surrounding greenery of the wet habitats where it grows. The flowers, while displaying an overall whitish green cast, are actually two-toned. The sepals are greener than the petals, lip, and spur. The hood formed by the dorsal sepal and petals is large compared to the size of the column. The lateral sepals are usually held out to the side, a feature that helps separate it from *P. aquilonis*, in which the lateral sepals usually are recurved. The lip of *P. huronensis* dilates slightly at the base, and the margins of the lower portion of the lip are slightly concave or acuminate. The lower margins of the rhombic lanceolate lip of *P. aquilonis* are slightly convex or even straight. Spur length and column shapes are also different from those on *P. aquilonis*. The spur of *P. huronensis* is about the same length as the lip to slightly longer, while the spur on *P. aquilonis* is shorter than the lip. The column in *P. aquilonis* is shorter than it is wide, and the anther sacs touch at the apex and diverge widely. The column on *P. huronensis* is taller than wide; the anther sacs do not touch at the apex and are nearly parallel.

Individual plants of *P. huronensis* may be relatively short-lived. Reddoch and Reddoch (1997) noted that individual plants appear to last for only 2 to 3 years, and indicated that this life span estimate took into account the irregular dormancy of a year or more, typical of many temperate zone terrestrial orchids.

There is often variation in the color of *P. huronensis*, especially within large populations, with some plants more nearly white than greenish. A few approach the bright whiteness of *P. dilatata* and may be the basis for *P. dilatata* being reported from New Mexico. Sheviak (1999b) proposed a possible reason for the whiter plants: *Platanthera huronensis* hybridizes with *P. dilatata*, and the plants with whiter flowers may be the result of ancient or recent gene flow between the species.

As with other *Platanthera* species, especially *P. aquilonis* and *P. dilatata*, *P. huronensis* presents newly opened flowers with the tip of the lip trapped in the hood formed by the petals and dorsal sepal. Catling and Catling (1991) believed that this ensures the removal of only one hemipollinarium per pollinator visit, since the upturned lip blocks access to the nectary directly from the front, forcing the insect to one side or the other.

According to Catling and Catling (1989), the pollinators of *P. huronensis* in Colorado included five species of bumblebees, three species of moths, and two species of butterflies. The insects tended to land in the middle of the spike and worked their way up before departing for another plant. Catling and Catling suggested that their data indicate that *P. huronensis* follows a generalized pollination strategy, with pollination by more than one family of insects, rather than a more specialized adaptation to a single species. Their study of a large group of blooming *P. huronensis* plants in Colorado revealed some to be self-pollinating because the pollinia had fallen forward onto the stigma, in the manner of *P. aquilonis*. Perhaps at this southern limit of its range, the pollinators of *P. huronensis* are less common and the plant self-pollinates as a survival strategy. Self-pollination also may simply be due to the close relationship with *P. acquilonis* or indicate gene flow from it. Reddoch and Reddoch (1997) reported self-pollinating flowers in Canada, so self-pollination may occur infrequently throughout its range.

Distribution

Platanthera huronensis is distributed from New York to Michigan, across Canada to Alaska, along the length of the Rocky Mountains and into New Mexico. Within New Mexico it occurs in the counties of Rio Arriba, San Miguel, and Taos. The locations in New Mexico represent its southern limit. It does not occur in Arizona.

Habitat

Platanthera huronensis grows in areas of constant moisture at elevations between 8000 and 10,200 feet (2440 and 3110 meters). The surrounding forest is usually mixed conifers and aspen. The most common habitat is in bright partial shade to full sun on hillside seeps and in wet meadows. It also grows on the edges of small streams, in marshy areas near larger streams and rivers, and in drainage ditches along roads and highways. It will sometimes grow in a ribbon of blooming plants around the margins of lakes and line the banks of streams feeding the lakes. Plants will colonize decaying logs in most of these environments. Like many of the *Platanthera* species, *P. huronensis* often develops extensive colonies and several hundred plants may be visible from one spot.

Blooming Season

Platanthera huronensis flowers in midsummer. The lowest flowers begin to open in the first week of July, and plants remain in bloom until early August. The seed capsules develop on the lowest flowers while the upper ones are still fresh and mature at the end of September and into early October.

Platanthera aquilonis blooms in the same area, and in the forest surrounding the open wet areas favored by *P. huronensis*, *Goodyera oblongifolia* and *G. repens* will be in spike. *Calypso bulbosa*, *Cypri-*

pedium parviflorum, *Listera cordata*, and all 4 *Corallorhiza* species will be in fruit.

Conservation

Platanthera huronensis is widespread and occurs in large colonies, often in slightly disturbed areas including man-made roadside ditches. Habitat destruction is always a threat, especially at the lower limits of its range in New Mexico, but significant habitat is protected within designated wilderness areas. Considered across its entire range, the status of *P. huronensis* is secure.

Notes and Comments

The various treatments of *P. huronensis* illustrate some of the historical confusion and different approaches regarding *Platanthera* since Lindley's description in 1835. Ames (1924) and Correll (1950) included *P. huronensis* within their definition of *H. hyperborea*. Farwell (1923) and Luer (1975) recognized some unique aspects and considered *P. huronensis* a variety of *H. hyperborea* and *P. hyperborea*, respectively (both now referable to *P. aquilonis*). Sheviak (1999a) pointed out that Rydberg's *Limnorchis media* and Luer's *P.* × *media*, the purported hybrid between *P. aquilonis* and *P. dilatata*, are actually synonyms of *P. huronensis*.

This treatment of *P. huronensis* follows that of Catling and Catling (1997), whose analysis and data supporting Lindley's original approach is the basis for again recognizing *P. huronensis* at the species level. They recommended that *P. huronensis* be treated as a species of hybrid origin, derived from *P. aquilonis* (*P. hyperborea* in their paper) and *P. dilatata*. In referring to *P. aquilonis*, *P. dilatata*, and *P. huronensis*, Catling and Catling concluded that "the three species are distinct without substantial backcrossing obscuring species boundaries." They showed *P. huronensis* to be intermediate between the two parents in terms of color, lip dilation, spur length, and anther separation. However, color alone was sufficient to separate the three species. Sheviak (1999b) concurred that *P. huronensis* is a species of hybrid origin, and that *P. dilatata* is one of the parents. However, he suggested that the other parent could be any one of several of the green *Platanthera* species.

Platanthera limosa Lindley
Annals Magazine Natural History 4: 381. 1840.

Etymology: The specific epithet is derived from a Latin word meaning "muddy," in reference to the wet and muddy habitat of this species.

Synonymy:
Habenaria thurberi A. Gray, Proceedings American Academy of Arts and Sciences 7: 389. 1868.
Habenaria limosa (Lindley) Hemsley, Biologia Centralli-Americana; Botany 3: 305. 1884.
Limnorchis arizonica Rydberg, Bulletin Torrey Botanical Club 28: 629. 1901.

Common names: Thurber's bog orchid.

Plate 23

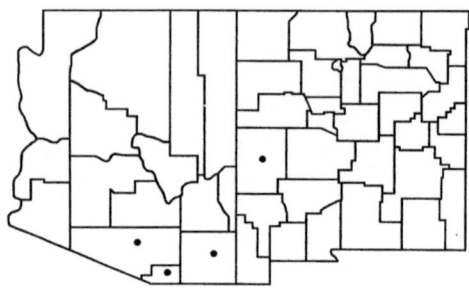

Map 27. Distribution of *Platanthera limosa*

Description

Plant: blooming plants from 20 to over 100 cm tall, reported to reach 1.5 meters tall in Mexico; densely flowered; largest plants with nearly 200 flowers on stem nearly 2 cm at base.

Roots: few fibrous roots on thickened rhizomes.

Leaves: five to nine, linear lanceolate, smooth and conduplicate, scattered along stem; lower leaves on largest plants over 40 cm long × 4 cm wide; more typically 20 to 30 cm long × 1 to 2 cm wide; becoming smaller higher on stem; uppermost leaves often bractlike; base of leaf sheathes the stem.

Floral bracts: narrowly lanceolate, larger than flower, 3 cm or longer on lowest flowers, becoming smaller on upper flowers; flowers often partially hidden by bracts.

Flowers: green to slightly yellowish green; 1.1 cm wide, 0.9 cm high with 1.5 cm spur; with musty odor.

Sepals: green; lateral sepals elliptic lanceolate, 5 mm long × 2 mm wide, twisting to hold large part of blade horizontal to ground; dorsal sepal elliptic ovate and slightly concave, 3.5 × 2.5 mm, forms hood with petals, hood held parallel to inflorescence axis.

Petals: yellow-green, ovate lanceolate, slightly falcate, form hood over column with dorsal sepal.

Lip: yellow-green, elliptic lanceolate to linear, 5.5 mm long × 1.5 mm wide, with small jagged callus.

Spur: yellow-green, filiform, two to three times as long as lip; 1.4 cm long × 1 mm wide.

Column: 2 mm high × 1.5 mm wide with parallel anther sacs filled with granular yellow pollen.

Capsule: ellipsoidal.

Platanthera limosa (lim-oh'-sa) is usually a densely flowered, robust plant. Elongated bracts along the inflorescence partially conceal the flowers. Smaller plants are often more laxly flowered, although the bracts are as well developed. The flowers are green to mostly yellow-green, with the lateral sepals held out to the side. The column fills less than half the hood formed by the dorsal sepal and petals. The anther sacs are parallel and widely spaced. The lip, elliptic lanceolate to almost linear, starts out at an oblique angle from the hood but bends sharply after a few millimeters to become vertical. There is a raised thickening along most of the lip, and a small jagged callus in the middle of lip at the bend. The spur is filiform, tapering at the apex, and about twice as long as the lip.

Identifying *P. limosa* in Arizona is fairly straightforward because it is the only *Platanthera* species in the Southeast corner of the state. In New Mexico identification is also rather easy, even though its range overlaps that of some other *Platanthera* species. *Platanthera limosa* can be distinguished by its long slender spur and the minute callus at the bend in the lip. Except for *P. brevifolia*, all the other *Platanthera* species in its range have short spurs. *Platanthera brevifolia* also has a long spur, and its range overlaps that of *P. limosa* in parts of New Mexico. The two are readily separated based on leaf shape. *Platanthera brevifolia* has attenuated bracts instead of fully developed leaves, while *P. limosa* has long narrow leaves scattered along the stem.

Distribution

Platanthera limosa is widely distributed in Mexico, including Baja California. In the United States it occurs only in the states of Arizona and New Mexico, where it reaches its northern range limit. In Arizona it is limited to the 3 southeastern counties of Pima, Cochise, and Santa Cruz. It is the only *Platanthera* species in these 3 counties, replacing *P. purpurascens*, which occurs just slightly to the north in Graham County. In New Mexico it occurs only in Catron County.

Habitat

Platanthera limosa typically grows in wet places in mixed forest at elevations between 6300 and 9150 feet (1920 and 2790 meters). The most common habitats are along a stream in light to bright shade, and on hillside seeps with a constant trickle of water at the roots. The hillside seeps are often steep, and footing can be treacherous owing to a jumble of fallen logs and small sinkholes hidden in the grasses. The plants are difficult to see because of the abundance of growth, and the leaves and even the flower spikes may be hidden in the mass of accompanying vegetation. In addition to many grasses and monkey flowers, the corn lily (*Veratrum californicum*) is a good indicator that the habitat is wet enough for *P. limosa*. Late in the season, the delightful blue gentian (*Gentiana grandis*) flowers in the same habitat. Less often, *P. limosa* grows in drier places at the bottom of drainages, but away from the stream, under shrubs or trees.

Blooming Season

Leaves of *P. limosa* emerge along streams and in seeps in early May. The plants develop slowly through May and June, and flowers begin to open the first week of July, about the beginning of the monsoon season. Peak blooming occurs in July and the first part of August, but a few flowers are often still in prime shape in late August or even early September. Plants in drier habitats do not put on as good a display as the wetter ones, and open up to 2 weeks later. The plants in dry areas do not bloom as fully and do not always develop completely. Scattered plants in dry habitats do not develop flowers at all, having only floral bracts.

Other orchids blooming in the same area as *P. limosa* include *Malaxis corymbosa, M. porphyrea, M. soulei,* and *M. abieticola. Corallorhiza maculata, C. striata, C. wisteriana,* and *Schiedeella arizonica* bloom earlier in more mesic habitats nearby, and are in fruit when *P. limosa* opens.

Conservation

Though fairly limited in distribution, *P. limosa* is not a conservation risk in Arizona. It grows in difficult-to-reach, often nearly inaccessible, habitat. Some of its sites are protected within designated wilderness areas and are therefore safe from any threat of development. In New Mexico, its situation is not as secure simply because of its relatively few sites there. It has not been confirmed in the state since 1903 and may have been extirpated. It is not on New Mexico's list of rare plants but belongs on List 1.

Platanthera purpurascens (Rydberg) Sheviak and Jennings
North American Native Orchid Journal 3 (4): 445. 1997.

Etymology: The epithet *purpurascens* refers to the intensely colored spots found on some flowers in parts of the range.

Synonymy:
Basionym: *Limnorchis purpurascens* Rydberg, Bulletin Torrey Botanical Club 28: 269. 1901.
Habenaria hyperborea (Linnaeus) R. Brown var. *purpurascens* (Rydberg) Ames, Orchidaceae 4: 90. 1910.
Platanthera hyperborea var. *purpurascens* (Rydberg) Luer, Native Orchids of the United States and Canada: 234. 1975.

Common names: none.

Plate 24

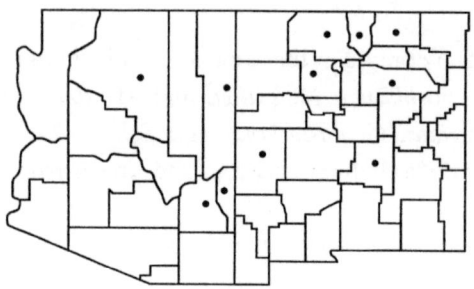

Map 28. Distribution of *Platanthera purpurascens*

Description

Plant: to over 90 cm tall, usually between 30 and 60 cm; three to six leaves, scattered along stem, spreading; laxly to densely flowered with 150 or more flowers on robust specimens.
Roots: few, short fibrous roots on thickened rhizome.
Leaves: lanceolate to slightly elliptic lanceolate; 10 to 15 cm long × 1 to 3 cm wide.
Floral bracts: lanceolate, to 2 cm long on lower flowers, shorter above.
Flowers: deep green, often with purple or reddish highlights.

Sepals: dorsal sepal ovate, 4 × 3 mm; lateral sepals elliptical lanceolate, 5 × 3 mm.
Petals: lanceolate, slightly falcate, 3.5 × 2 mm.
Lip: linear lanceolate, strongly dilated at the base, 6 × 3 mm.
Spur: scrotiform to clavate, shorter than the lip, 3 mm.
Column: small, about as broad as high, less than half as large as hood formed by dorsal sepal and petals, anther sacs separated at apex, diverging toward viscidia to nearly parallel.
Capsule: ellipsoidal.

Platanthera purpurascens (pur-pur-as'-enz) is easily identified by its aroma. Sheviak and Jennings (1997) described the strong musty smell as characteristic of the species. Others in the genus such as *P. aquilonis* either have no aroma or like *P. dilatata*, have a sweet spicy aroma. Flower presentation and plant habit offer additional clues to identification. The slender inflorescence is loose to moderately dense, and the flowers are sometimes arrayed in fascicles. The leaves diverge widely from the stem and are blunt, instead of fairly acute as on most other *Platanthera* species in this area. Sheviak (1999a) noted that the wide-spreading leaves are a reliable character because they do not vary with exposure to sun or amount of shade. The leaves are also short compared to robust specimens of *P. huronensis*, but leaf size is not a reliable determinant because of overlap with smaller plants of other *Platanthera* species.

Intense red to purplish blotches on the flowers from which the name derives are not evident in the region under study. Some plants do have markings on the flowers that with some imagination can be construed to be colored blotches but may just be environmental blemishes. Sepals and petals are a deep green. Lip color varies from yellowish green to deep green, with what Sheviak (1999a) described as an intense, translucent, watery quality, which he said is characteristic of the species. Lip shape varies from linear elliptic to lanceolate from a dilated base. The spur is always much shorter than the lip, but spur shape is quite variable, ranging from clavate to strongly saccate or scrotiform. Extremely short scrotiform spurs resemble those of *P. stricta* and are the primary reason the 2 species are often confused.

Lip shape and spur length and shape are extremely variable in this species and may be a result of hybridizing with other *Platanthera* species. In the herbarium and in the field, the flowers are often confused with those of *P. aquilonis* and *P. stricta*, and the spur shape on *P. purpurascens* covers the full range of shapes and lengths on those 2 species. It well may be that there are at least two taxa currently masquerading as *P. purpurascens*, but for now it is treated as the most variable species in the complex.

Distribution

Platanthera purpurascens is concentrated in the Rocky Mountain states of Colorado and New Mexico, with populations in adjacent Arizona. Coleman (1995) reported disjunct locations in California, where it is relatively rare. Within Arizona it occurs in the 4 counties of Apache, Coconino, Graham, and Greenlee. It grows in Catron, Colfax, Lincoln, Rio Arriba, Sandoval, San Miguel, and Taos Counties in New Mexico. The southern range limit for *P. purpurascens* is on Mt. Graham in Graham County, Arizona.

Habitat

Platanthera purpurascens grows in damp to wet places at elevations between 7000 and 10,240 feet (2130 and 3125 meters). It shows a wide tolerance for habitat. Primary habitats include riparian areas, seeps, and meadows in partial shade to full sun part of the day. Along streams and in seeps, it grows at the base of rocks and on wet banks. Companion plants include grasses, monkey flowers (*Mimulus* species), and corn lily (*Veratrum californicum*). As does *P. brevifolia*, *P. purpurascens* also inhabits dry sites, in partial to moderate shade. The plants growing in relatively dry conditions are limited to the northern halves of these two states.

Blooming Season

Platanthera purpurascens emerges in late May to early June, and at this stage the shiny green leaves are easily detected. Later in the season the leaves may be totally obscured, with only the inflorescence appearing above surrounding growth. Flowers start to open about the second week of July and open successively up the stem. Some blossoms remain in good condition until past the middle of August, and a few upper flowers may still be fresh into early September. Seed capsules dehisce in late September and early October, and soon afterward aboveground growth is killed back by frost.

Malaxis soulei often blooms in the same area and at the same time as *P. purpurascens*. In the surrounding forest, *Goodyera oblongifolia* and *G. repens* will certainly be in spike, and late in the season a few flowers may be open. Earlier in the season, *Corallorhiza maculata*, *C. striata*, *C. wisteriana*, *Schiedeella arizonica*, and *Cypripedium parviflorum* may have bloomed nearby.

Conservation

Platanthera purpurascens is widely distributed and grows in diverse habitats, and so is not threatened when considered across its range. Portions of its habitat are protected within wilderness areas. Individual plants are subject to browsing by deer, elk, and cattle, and sometimes most plants in an area are eaten.

Notes and Comments

Rydberg (1901) first described *P. purpurascens* as a species in his genus *Limnorchis*, which he established to segregate from *Platanthera* (or *Habenaria*) a group of species related to *P. dilatata* and *P. hyperborea*. Most authors since Rydberg merged *Limnorchis purpurascens* within other taxa. Coulter (1909) may have been the first to treat *P. purpuras-*

cens as synonymous with *P. stricta*. Ames (1924) treated it as a variety of *H. hyperborea*, while Correll (1950) considered it a synonym of *H. saccata*. Most regional floras subsequent to Correll's work, such as that by Kearney and Peebles (1951), followed his approach. Schrenk (1978) suggested that *P. purpurascens* may be a hybrid between *P. hyperborea* and *P. stricta*, which he called *P.* × *correllii*. There were two notable exceptions. Tidestrom and Kittell (1941) recognized it as *Habenaria purpurascens*, and Luer (1975) followed Ames in calling it *P. hyperborea* var. *purpurascens*. Recent work by Sheviak and Jennings (1997) convinced them of the soundness of Rydberg's original approach. They give credit to Luer for once again recognizing the significance of this taxon, and they transferred *L. purpurascens* to *Platanthera*.

Distribution maps in the book by Luer (1975) show *P. purpurascens* in the Rocky Mountains in Colorado and New Mexico. It also is in Arizona, but that is not evident from Luer's map. Luer shows *P. stricta* in the northwestern United States and Canada. Their ranges do not overlap, and *P. stricta* does not occur in either Arizona or New Mexico.

Some of the extreme variation within *P. purpurascens* is most likely due to recent or ancient hybridization. Sheviak (1999a, 2000) believes that some of the plants in the Southwest are hybrids between *P. purpurascens* and *P. dilatata* var. *albiflora*.

Platanthera sparsiflora (S. Watson) Schlechter
Bulletin L'Herbier Boissier 7: 538. 1899.

Etymology: *Sparsiflora* is from Latin and means "scattered flowers" or "sparsely flowered," in reference to the supposedly laxly flowered spikes.

Synonymy:
Basionym: *Habenaria sparsiflora* S. Watson, Proceedings American Academy of Arts and Sciences 11: 276. 1877.
Limnorchis sparsiflora (S. Watson) Rydberg, Bulletin Torrey Botanical Club 28: 631. 1901.
Habenaria aggregata Howell, Flora of Northwest America: 628. 1902.
Limnorchis aggregata (Howell) Frye and Rigg, Northwest Flora: 114. 1912.
Habenaria leucostachys (Lindley) S. Watson var. *viridis* Jepson, Flora of California: 331. 1921.

Common name: sparsely flowered bog orchid.

Plate 25

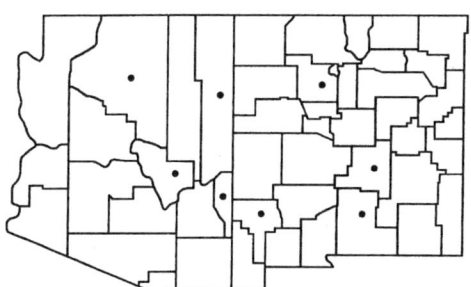

Map 29. Distribution of *Platanthera sparsiflora*

Description

Plant: from 8 to 60 cm tall here and to nearly 100 cm elsewhere; 30 to over 100 green flowers; three to five alternating leaves replaced by bracts higher on stem.
Roots: a few short, 2-mm-thick roots.

Leaves: linear to elliptic lanceolate, to 11 cm long and 2 cm wide.

Floral bracts: lanceolate, nearly 2 cm long at lowest flowers, smaller above.

Flowers: green to yellowish green, 1.7 cm tall × 0.3 cm wide; petals and dorsal sepal form hood over column, lateral sepals usually wrap behind spur or held to side.

Sepals: lateral sepals light green, linear, falcate, 6.5 mm long × 2 mm wide; dorsal sepal dark, dull green, ovate, 5 mm long × 3 mm wide.

Petals: pale green, ovate lanceolate, falcate, 6 mm long × 2 mm wide.

Lip: pale green to yellowish green, linear to lanceolate, acute, with central thickening at column, 7 mm long × 1.5 mm wide.

Column: green, occupies half or more of the hood formed by petals and dorsal sepal, anther sacs widely separated at top, diverging toward viscidia; 2.5 mm high × 2 mm wide; pollinia granular, yellow.

Spur: pale green, cylindrical, curved, 1 cm long × 1 mm thick.

Capsule: ellipsoidal, 10 mm long, 4 to 5 mm wide, held upright.

Platanthera sparsiflora (spar-si-flor'-ah) is called the sparsely flowered bog orchid, but that common name is somewhat of a misnomer because mature plants often have more than 100 green to very yellowish green flowers. It is most easily recognized owing to a large column that fills half or more of the hood formed by the dorsal sepal and petals. Usually the lateral sepals are reflexed and sometimes are partially twisted, giving the flowers a narrow profile. The lip, which varies in length, is linear though variable in width and has a short thickening at the base. The spur is also variable in length, from slightly shorter than the lip to as much as 1.5 times longer than the lip. The flowers have a strong aroma. The ovaries angle out from the inflorescence axis, resulting in a slight downward cast to the flowers.

Plant habit is variable. Typical stems support evenly spaced lanceolate leaves, but some plants have longer lanceolate leaves clustered near the bottom of the stem. Luer (1975) called those plants *P. sparsiflora* var. *ensifolia* (Rydberg) Luer. Plants meeting the description of *P. sparsiflora* var. *ensifolia* occur in both Arizona and New Mexico.

Distribution

Watson (1877) described *P. sparsiflora* from material collected in the Sierra Nevada of northern California. It grows in much of the western United States and in parts of Mexico. In the United States it ranges from Washington, Oregon, and California to Utah, Arizona, and New Mexico. In Arizona it is in the counties of Apache, Coconino, Gila, and Greenlee. In New Mexico it is in Grant, Lincoln, Otero, and Sandoval Counties.

Habitat

Platanthera sparsiflora grows at elevations between 5200 and 8500 feet (1980 and 2590 meters) in these states, although it gets to an elevation of at least 11,000 feet (3350 meters) in California. It requires a constantly wet substrate to survive, and so is limited to areas with a permanent source of water. A typical habitat is a large seep on the side of a canyon. The ground is usually rocky and totally saturated, often with a flow of surface water, and footing is slippery and treacherous. The plants nestle among the rocks, often in full sun or partially protected by small shrubs. Other habitats are along the sides of small streams and in damp portions of meadows in full or nearly full sun. It colonizes manmade conditions that duplicate its natural habitats, such as roadside ditches and the edges of culverts.

Blooming Season

The blooming season for the sparsely flowered bog orchid stretches from the middle of May to the first of August, but peak bloom is during July. The flowering season lasts later in California where *P. sparsiflora* is more common and can often be found with still fresh flowers in early September. *Corallorhiza maculata* and *C. striata* bloom in drier areas nearby, as does *Goodyera oblongifolia*, though later in the season. In one part of its range, it blooms with *P. zothecina*, and intermediate forms suggest the two are hybridizing.

Conservation

The sparsely flowered bog orchid exists in sufficient numbers in widely scattered locations to be safe from immediate threats. Significant portions of its habitat are protected within wilderness areas. Cattle eat the plants and in areas of heavy grazing, none of the plants may survive to bloom.

Platanthera zothecina (Higgins and Welsh) Kartesz & Gandhi
Phytologia 69(3): 134. 1990.

Etymology: The name refers to its usual location in a recess or niche on a canyon wall.

Synonymy:
Basionym: *Habenaria zothecina* Higgins and Welsh, Great Basin Naturalist 46(2): 259. 1986.
Limnorchis zothecina (Higgins and Welsh) W. A. Weber, Phytologia 67(6): 427. 1989.
Platanthera zothecina (Higgins and Welsh) Catling and Sheviak, Lindleyana 8(2): 81. 1993.

Common name: alcove bog orchid.

Plate 26

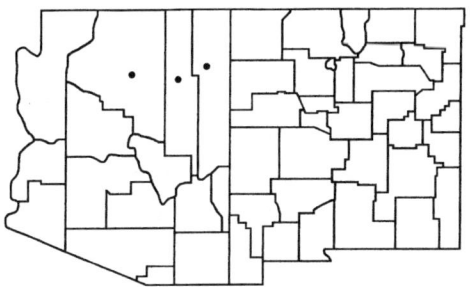

Map 30. Distribution of *Platanthera zothecina*

Description

Plant: to 35 cm tall with four or five mostly basal leaves replaced by bracts higher on stem; one or two bracts between leaves and flowers; 5 to 30 green flowers, usually laxly spaced.

Roots: a few thick roots.

Leaves: four or five, lower leaf very rounded, others more oblanceolate to ovate elliptic, 3 × 2 cm to 18 × 6 cm.

Floral bracts: lanceolate.

Flowers: greenish to yellowish green.

Sepals: green; dorsal sepal broadly ovate, 4 to 5 mm × 3 to 4 mm; lateral sepals broadly ovate lanceolate, slightly falcate, 6.5 × 2 mm.

Petals: green, oblanceolate, falcate, 5 × 2 mm.

Lip: linear to linear elliptic, 0.8 to 1.0 cm with widest part near center; green with yellowish highlights on raised horizontal bump across lip near nectary, but bump may not be on uppermost flowers.

Spur: much longer than lip, 1.3 to 2.6 cm long, slightly clavate, lower half curves away from lip.

Column: large, green with yellowish highlights, anther sacs separated at top, nearly parallel but spreading at viscidia.

Capsule: ellipsoidal.

Platanthera zothecina (zoth-e-ki'-na) is a relatively recently described western member of the genus. Higgins and Welsh (1986) described it based on a specimen from Grand County, Utah. Prior to their description, it had been included within *P. sparsiflora*, although it was often considered a robust form of that species. Emerging plants appear quite distinct from *P. sparsiflora* because of their much more rounded leaves. Leaves on mature plants of *P. zothecina* are clustered at the base of the stem, with the lower one or two nearly ovate. The flowers are green to yellowish green, and the column is large compared to the hood formed by the dorsal sepal and petals. The flowers can be distinguished from *P. sparsiflora* by the shape of the lip and length of the spur. The lip of *P. zothecina* is usually linear elliptic and has a distinct horizontal ridge across it about one-third the length of the lip down from the junction with the column. The lip bends downward at this ridge. The lip of *P. sparsiflora* is linear, with a short vertical ridge, and does not bend. The spur in *P. zothecina* is longer than the spur in *P. sparsiflora*, from 1.5 to 2.5 times the length of the lip. In *P. sparsiflora* the spur ranges from slightly shorter than the lip to 1.5 times as long. Spur lengths are referenced to the lowest flowers on the stem, and to those that have been opened several days. In *P. zothecina*, the spurs lengthen for several days after the flowers open, and spurs are not as long on upper flowers.

Distribution

Platanthera zothecina is mainly confined to the Four Corners area of the southwestern United States — the region where the states of Arizona, New Mexico Utah, and Colorado share a common point. It has not been reported from New Mexico but should be expected there because of its proximity in adjacent Arizona. In Arizona, *P. zothecina* has been found only in Apache, Coconino, and Navajo Counties. The discovery in Apache County was made by D. Roth and D. Mikesic in 1999. Most of the plants occur in the drainage of the upper Colorado River north of Interstate Highway I-40 and east of Flagstaff, but a few are along the Grand Canyon west of Flagstaff, and one colony south of Flagstaff is the southern limit for *P. zothecina*.

Habitat

Platanthera zothecina grows at elevations between 5000 and 9000 feet (1525 and 2759 meters). The common name of alcove bog orchid indicates its usual habitat. It grows in damp to wet alcoves on the walls of sandstone canyons prevalent in the Four Corners region. Native Americans sometimes selected larger alcoves as sites for cliff-dwelling homes. Some of these sites have water sources such as small streams or seeps within the alcove, providing moisture for the colonies of *Platanthera zothecina*. In one of the larger alcoves the plants are so protected by the overhanging rock that they receive direct sun only a portion of the day, and little if any direct rain, relying wholly on groundwater.

Seeps and the banks of small streams in the canyon bottoms are prime habitats for *P. zothecina*. The seeps are perhaps the more picturesque of the habitats of *P. zothecina* because companion plants often bloom with the orchid. In late June the seeps can be a riot of color due to yellow columbine (*Aquilegia chrysantha*), red and yellow monkey flower (*Mimulus cardinalis* and *M. guttatus*), and purple monks' hood (*Aconitum columbianum*) cascading down the rocky slopes, all but hiding the orchids.

Seeps high on the walls of canyons are called *hanging gardens* and appear as bright green spots in an otherwise nearly barren landscape. Many of the hanging gardens are tantalizingly out of reach and can be explored only with binoculars. Others are accessible after a slight climb, although the dripping water, running across sandstone worn smooth by centuries of wind and water erosion, makes climbing perilous.

Blooming Season

Leaves of *P. zothecina* appear in late April to early May. Flower spikes develop during May and early June, and flowering commences in mid-June, lasting to the first of August. The flowering period at a given location lasts about 5 weeks. Seed capsules mature about 2 months after the flowers first open. *Epipactis gigantea* is a frequent flowering companion in the hanging gardens and seeps. *Platanthera sparsiflora* blooms near *P. zothecina* in the southern part of its range, and plants of intermediate form occur, suggesting hybridization is taking place.

Conservation

The status of *P. zothecina* is not yet completely understood, primarily owing to its recent description, but the herbarium records suggest it is rare throughout its range. The 11 herbarium records from Arizona document only six locations, discovered between 1933 and 1991. Recently botanists with the Colorado Plateau Research Station and with the Navajo Natural Heritage Program have pinpointed several additional sites. Several of its historical locations are protected within the Grand Canyon National Park and Navajo National Monument. Most of the recently discovered locations are on land belonging to the Navajo Nation. Though rare, *P. zothecina* is being studied actively, and many of its known sites are safe from development. It is ranked G2/S1 in Arizona based on the relatively few acres occupied rather than strictly on the number of occurrences.

Notes and Comments

Several sites of *P. zothecina* are located within National Park and National Monument boundaries. In order to gather data necessary to understand the conservation needs of this rare orchid, in the summer of 1999 the National Park Service initiated a study of *P. zothecina* within the Navajo National Monument in northern Arizona. Four groups of plants were staked, counted, and monitored throughout the growing season. The study, destined to last several years, should yield data on population dynamics and flowering patterns.

Table 9. *Platanthera* Characteristics

Species	Leaves	Column	Lateral Sepals
P. aquilonis	Ascending, scattered along stem, oblong to linear lanceolate	Small, wider than high; anther sacs touching, widely diverging toward viscidia; self-pollinating with pollinia rotating forward and fragmenting on stigma	Reflexed down along spur
P. brevifolia	Reduced to bracts, erect, sheathing stem	Large, anther sacs nearly parallel to slightly spreading at viscidia	Widely spreading
P. huronensis	Ascending to arcuate spreading (bowed), scattered along stem, oblong to linear lanceolate	Small, but larger than P. aquilonis, narrower than high, anther sacs almost parallel to slightly diverging toward bottom	Spreading to slightly reflexed and twisted
P. limosa	Spreading, ascending, linear lanceolate	Small, slightly taller than wide; anther sacs parallel, widely spaced at top and bottom	Widely spreading
P. purpurascens	Blunt, short, scattered along stem, abruptly spreading, often right angle to stem	Small, as broad as high, anther sacs separated, diverging toward viscidia, connective clearly visible	Reflexed, twisted
P. sparsiflora	Linear to elliptic, lanceolate scattered along stem; sometimes mostly basal	Large, broad; anther sacs widely separated at top, diverging toward viscidia	Reflexed and partially twisted to slightly spreading
P. zothecina	Broad, oblanceolate to ovate elliptic, mostly basal	Large, broad; anther sacs separated at top, nearly parallel, but spreading near viscidia	Sharply reflexed

Table 9. Continued

Species	Lip	Spur	Inflorescence
P. aquilonis	Rhombic lanceolate to lanceolate, not acuminate, base not rounded or dilated, yellow green	Shorter than lip, clavate to cylindrical, curved, blunt	Lax to dense, scentless
P. brevifolia	Lanceolate to elliptic lanceolate	Slenderly cylindrical to filiform, apex tapered, much longer than lip	Dense to lax
P. huronensis	Acuminate, lanceolate to linear lanceolate, base slightly rounded and dilated, whitish green, strong sweet odor	Slenderly cylindrical to barely clavate	Lax to dense, whitish green, sweet scent
P. limosa	Elliptic lanceolate to linear, yellow green, median basal thickening; small callus	Filiform, tapering toward apex, much longer than lip	Lax to dense
P. purpurascens	Linear lanceolate, strongly dilated at base	Scrotiform to saccate or strongly clavate, shorter than lip	Lax to dense, flowers often grouped in fascicles at irregular intervals, strong musty aroma
P. sparsiflora	Linear to lanceolate, apex acute	Slightly shorter than lip to longer than lip, 0.7 to 1.4 cm	Lax to dense; flowers with strong aroma
P. zothecina	Linear to linear elliptic, middle thickened toward base	Much longer than lip, 1.3 to 2.6 cm	Lax

Schiedeella Schlechter

Beihefte zum Botanischen Centralblatt 37 (2): 379. 1920.
Etymology: The genus was named after Christian J. W. Schiede, a German naturalist who studied and collected Mexican plants.

Schiedeella (sheed'-ee-el-ah) was first proposed by Schlechter in 1920 to provide for plants he separated from *Spiranthes* based on the shapes of the column, rostellum, and stigma. Prior to Schlechter's treatment, the plants were included in other genera within the subfamily Spiranthinae. Schlechter's work was not widely accepted, however, and many authors such as Correll (1950), Ames and Correll (1952), Williams (1965), and Luer (1975) continued to treat certain members of the genus as *Spiranthes*. This situation began to change starting in the 1980s when both Balogh (1982) and Garay (1982) recognized *Schiedeella*, although they had significant differences in their approaches. Szlachetko (1992), in an extensive revision of *Schiedeella*, retained 10 species in two subgenera. Six species he placed in the subgenus *Schiedeella*, and 4 species he placed in a new subgenus he named *Schiedeellopsis* Szlachetko. Dressler (1993) also recognized 10 species. *Schiedeella* is found primarily in Central America, the Greater Antilles, and Mexico. One species occurs in the United States, in Arizona, New Mexico, and Texas.

Schiedeella arizonica P. M. Brown
North American Native Orchid Journal 6 (1): 3. 2000.

Etymology: The species is named for the state in which the plants were first observed.

Synonymy: none.

Common name: fallen ladies'-tresses.

Plates 27, 28

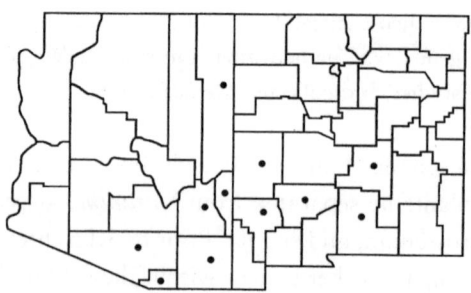

Map 31. Distribution of *Schiedeella arizonica*

Description

    Plant: leafless in bloom to 33 cm tall, but typically 10 to 20 cm tall; stems green to tan to light rose; two or three leaves in basal rosette appearing after flowering; 3 to 14 flowers on top third of stem.
    Roots: spheroidal, cormlike rhizome, with a few fibrous roots.
    Leaves: simple, ovate elliptic, 1.5 × 6 cm.
    Floral bracts: lanceolate, appearing dried or translucent at flowering.
    Flowers: whitish, rose, or tan; 6 mm long × 4 mm wide × 4 mm high; petals and dorsal sepal form hood over lip, resulting in narrow tubular flower; sepals and petals so thin as to be translucent.
    Sepals: whitish, rose, or tan; lateral sepal linear, 6 mm long × 1.5 mm wide, outer third curls inward to touch lip; dorsal sepal linear to very slightly elliptic linear.

Petals: whitish, rose, or tan; spatulate to oblanceolate, slightly falcate; 4 mm long × 1.5 mm wide.

Lip: whitish, rose, or tan; 5 to 6 mm long × 4 mm wide; oblong, nearly quadrate, constricted above the middle and near the apex; narrows to a slight claw near the column; apex crenulate; thickened in the center with red to red-orange spot; three green stripes run length of lip, with additional stripe at each side for basal two-thirds; center portion of lip finely pubescent; tip of lip curls back completely.

Column: green, 4 mm long × 1.5 mm wide; two pairs of yellow pollinia, each pair banana-shaped; brown anther cap; stigma dark green.

Capsule: upright, ellipsoidal, 8 × 2 mm.

The dainty solitary flower stem of *Schiedeella arizonica* (ar'i-zo'-ni-ca) rises from a carpet of pine needles, looking like a miniature mycotrophic plant and resembling small specimens of *Corallorhiza wisteriana*. The typical tan to rose color of the spike and the total absence of leaves from emergence through anthesis led to the original assumption that the plant was parasitic.

Schiedeella arizonica is a delightful miniature. The flowering stem looks not much different from a pine needle on the forest floor, and a magnifying glass is needed for careful study of the tubular flowers. The characteristic red to red-orange spot covering the center half of the lip is best viewed from the underside of the flower. The dorsal sepal and petals form a hood over the lip, and the lateral sepals curve around and contact the front of the lip, which curves backward on itself. Two color forms grow side by side in most areas. The dominant form is cream to tan, with a slight rose suffusion in the lip, sepals, and petals. The stem, bracts, sepals, petals, and lip are the same basic color. Five green stripes on the lip, three of which run the full length, add some additional color. Some plants, especially at the northern limit of the range, have stems and flowers of a solid pale pink-rose color. The second, less numerous, form is probably anthocyanin-free and has green stems and bracts with whitish flowers. The rose suffusion is absent, and the green stripes create a striking contrast on the lip.

Though thought to be parasitic at the time of its discovery, *S. arizonica* is photosynthetic. The small basal rosette of leaves appears shortly after the capsules dehisce, about the beginning of the monsoon season in midsummer. There are two to five leaves per rosette, but the rosettes sometimes grow in clusters, so they appear denser than they are. The leaves persist until fall, when they die back for the winter.

Distribution

Schiedeella arizonica is widely distributed in the southwestern United States in Arizona, New Mexico, and Texas. P. M. Brown (2000) stated that the extent of its range is not totally understood, but he could not find any herbarium records of it from Mexico. Ames and Correll (1952) and McVaugh (1985) gave the range of *Spiranthes parasitica* as encompassing the southwestern United States to Central America. They were discussing the range of the taxon as understood then, and did not have the benefit of the scholarship of Balogh (1982), Garay (1982), and Szlachetko (1992), so their ranges may not be representative of *Schiedeella arizonica*. It clearly should be sought in mountainous regions of Mexico immediately south of New Mexico and Arizona. In Arizona it occurs in 6 counties: Apache, Cochise, Graham, Greenlee, Pima, and Santa Cruz. In New Mexico it is in the counties of Catron, Grant, Lincoln, Otero, and Sierra. It reaches the northern limits of its range in the southern part of Apache County, Arizona, and Catron County, New Mexico.

Habitat

Schiedeella arizonica grows in mesic, mixed coniferous-deciduous forest at elevations between 6450 and 9300 feet (1960 and 2830 meters). On some slopes it grows among rocks or in bare soil, but usually it is in heavy forest duff in flat to very steep terrain. Less often, it ventures out onto the edges of meadows. Common companion herbaceous plants include lupines, violets, and ferns.

Blooming Season

The leafless flower spikes of *S. arizonica* emerge in late April and mature into early May. Flowering lasts from about the second week of May until late June, but peak bloom is from the last week of May into the second week of June. Seed capsules mature and dehisce within 4 to 5 weeks. Leaves sprout within 6 weeks after blooming, with non-blooming plants putting up leaves first. Only about 10 to 15 percent of plants bloom in any given year, and only a few bloom several years in a row. The low percentage of flowering among *Schiedeella* species is not unique to *S. arizonica*. Szlachetko (1993), describing *S. romeroana* as a new species, quoted the collection notes with the type specimen taken by Greenwood as stating that only 10 of 200 plants produced flowers.

Many other orchids grow in the same habitat as *S. arizonica*. It blooms the same time as, and within inches of, *Corallorhiza wisteriana*, and late in its season, may bloom with early plants of *C. maculata* and *C. striata*. At higher elevations it grows with *Platanthera purpurascens* and *Goodyera oblongifolia*. In southeastern Arizona, it may grow intermixed with *Malaxis corymbosa*, *M. porphyrea*, *M. soulei*, and *M. abieticola*, although the *Malaxis* species do not bloom until long after the leaves of *S. arizonica* appear.

Conservation

Long believed to be rare in the United States, *S. arizonica* is actually plentiful in Arizona and New Mexico. Although unknown until 1906 when John Thornber found it in the Santa Catalina Mountains near Tucson, Arizona, it is widely scattered and locally common, and incidentally, still blooms in the Santa Catalina Mountains. Its presumed rarity is due more to its flowering habits than lack of numbers. The fallen ladies'-tresses is the most difficult orchid to notice in Arizona and New Mexico because its small size and habitat-matching colors camouflage it perfectly among debris on the forest floor. It is so well hidden that a common experience of orchid hunters is to finally locate a blooming plant, only to lose it after a brief glance away. It may take several moments to relocate it, even when standing within a few feet of the plant. Another reason it was thought to be rare is its low bloom rate. Since only a small percentage of plants bloom each year, population estimates based on blooming plants may be low by an order of magnitude. The best way to find flowering plants of S. *arizonica* is to look for leaves in August when they are plentiful and easy to identify. Simply return the following spring and search the same area for flowering plants.

Notes and Comments

Schiedeella arizonica is the name now applied to the southwestern taxon that for many years was referred to as either *Schiedeella parasitica* (Richard and Galeotti) Schlechter or *Spiranthes parasitica* Richard and Galeotti. Szlachetko (1992) pointed out that this was due to an error somewhere along the way, and that Richard and Galeotti never saw the plant that for so long would carry the name they assigned it.

It took some indirect detective work to determine that the taxon in the southwestern United States needed a new name. In 1978 Donald Dod described a new species from the Dominican Republic he called *Spiranthes fauci-sanguinea*. Burns-Balogh (1989) recognized it as a *Schiedeella* and transferred it to that genus. She observed that the type specimen of *Spiranthes parasitica* had several significant differences from the southwestern plants that had been going by that name. She specifically pointed out that the specimen that was identified as *Spiranthes parasitica* in the photograph from Pima County, Arizona, in the publication by Luer (1975) was actually *Schiedeella fauci-sanguinea*. Szlachetko (1992, 1993) supported Burns-Balogh's position, stating that the name *S. parasitica* was a synonym for *Schiedeella violacea*. Burns-Balogh and Szlachetko demonstrated that the name *S. parasitica* did not apply to the plants in the southwestern United States. However, there remained doubt among some botanists that *Schiedeella fauci-sanguinea* was the proper name for those plants. P. M. Brown (2000) addressed that issue when he pointed out morphological differences between the southwestern plants and *Schiedeella fauci-sanguinea*. Brown could find no other validly published name that applied to the southwestern taxon, and named it in honor of the state of the original discovery.

The rules of botanical nomenclature must be followed to maintain order, and much effort is expended sorting out the correct name for a taxon. Sometimes, though, a certain beauty and elegance is lost when a name long used is proved to be incorrect. Such is the case with *S. arizonica*. The incorrectly applied name of *S. parasitica* was poetically appropriate, though always a misnomer, because of its growth habit.

Spiranthes L. C. Richard

Mémoires du Museum d'Histoire Naturelle Paris 4: 50. 1818.

Etymology: *Spiranthes* is from two Greek words meaning "coil" and "flowers," referring to the coiled or spiraled flower spikes of the genus.

Spiranthes (spy-ran'-theez) is a member of a large and confusing subfamily. In their respective revisions of the Spiranthinae, Garay (1982) identified 390 species in 44 genera, and Dressler (1993) identified 409 species in 41 genera. Garay recognized 42 species of *Spiranthes* scattered throughout much of the Northern and Southern Hemispheres, and Dressler recognized 30 species. About 20 *Spiranthes* species occur in the United States and Canada, excluding Florida. In an apparent allusion to the resemblance of the floral spirals to certain hairstyles, *Spiranthes* are commonly called ladies'-tresses.

Luer (1975) included *Schiedeella* Schlechter and *Stenorrhynchos* Sprengel within his expanded concept of *Spiranthes*. Most floral treatments of Arizona and New Mexico, such as the works by Tidestrom and Kittell (1941), Kearney and Peebles (1951), and Lehr (1978), like Luer, included *Schiedeella* and *Stenorrhynchos* as part of *Spiranthes*. Both Garay (1982) and Dressler (1993), following earlier works, once again recognized them as separate genera. The differences among the three genera are sufficient to maintain them separately, and therefore, only 3 *Spiranthes* species are recognized in Arizona and New Mexico, with 2 species formerly included within *Spiranthes* treated as either *Schiedeella* or *Stenorrhynchos*.

Three species of *Spiranthes* grow in Arizona and New Mexico, but one other potentially grows there based on the existence of proper habitat and the presence of colonies in a neighboring state. *Spiranthes diluvialis* Sheviak occurs in southern Utah and should be looked for along river and stream systems that enter Arizona from Utah. The 3 *Spiranthes* species that are found in Arizona and New Mexico can be separated easily based on a combination of geography and elevation. In Arizona, the only *Spiranthes* species south of Interstate Highway I-10 is *S. delitescens*. In New Mexico, the *Spiranthes* species found north of I-10 and below 6000 feet (1830 meters) is *S. magnicamporum*. In either state, the *Spiranthes* species found above 8000 feet (2440 meters) is *S. romanzoffiana*. Floral features also aid in identification, as shown in the key below. When you use the key, apply it to fully opened flowers on the lower portion of the stem. Some of the important features such as lip shape tend to be less distinct in flowers higher on the stem.

Key to the Species of *Spiranthes*

1. Lip pandurate, the apex dilated, sepals and petals united throughout their lengths and forming a hood   *S. romanzoffiana*
1a. Lip ovate to oblong, the apex only slightly or not at all dilated, sepals and petals with apices free and spreading
    2. Lip oblong, rachis with obscurely glandular hairs   *S. delitescens*
    2a. Lip ovate, rachis with glandular hairs   *S. magnicamporum*

Spiranthes delitescens Sheviak
Rhodora 92: 215. 1990.

Etymology: The specific epithet is based on the present participle of *delitescere*, meaning "to hide away," in reference to the species' relatively recent discovery.

Synonymy: none.

Common name: Canelo Hills ladies'-tresses; Madrean ladies'-tresses.

Plate 29

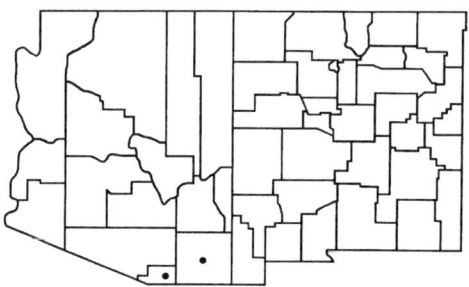

Map 32. Distribution of *Spiranthes delitescens*

Description

Plant: 20 to nearly 60 cm tall, leaves on lower half of stem, multiple bracts above leaves to flowers; fine hairs on upper portion of stem; up to 45 flowers on loose to dense spike, not always clearly spiraled.

Roots: three to eight, thick, fibrous, 5 cm long and 0.5 cm thick.

Leaves: three to five, linear to lanceolate; to 18 cm long × 1.5 cm wide, usually persisting past flowering, but sometimes fading at flowering.

Floral bracts: elliptic lanceolate, 8 to 12 mm long, decreasing in size higher on stem.

Flowers: white, tubular, 1.0 cm long × 0.6 cm wide × 0.6 cm high, with lateral sepals free and spreading and dorsal sepals and petals forming hood over column.

Sepals: white, with fine hairs on outside; dorsal sepal ovate elliptic, to 9 mm long × 2 mm wide; lateral sepals ovate lanceolate, to 8 mm long × 2 mm wide.

Petals: white, lanceolate, held close to dorsal sepal, to 8 mm long × 1.5 mm wide.

Lip: white with yellowish or greenish tinge in center, ovate with expanded crenulate apex, to 8 mm long × 5 mm wide, apex turns under and curves back.

Column: slender, 3 mm high × 1.5 mm wide; two powdery, mealy yellow pollinia.

Capsule: ellipsoidal, to 9 mm long × 4 mm wide, covered with fine hairs.

Spiranthes delitescens (del-i-tes'-ens) is the only orchid in either Arizona or New Mexico listed as an endangered species. The plant is tall and slender, with a relatively small portion of the total inflorescence covered with flowers. This feature helps position the flowers above surrounding grasses and makes them more visible to pollinators. Sheviak (1990) differentiated *S. delitescens* from the somewhat similar *S. diluvialis* and *S. graminea* based on floral characteristics. The flowers of *S. delitescens* appear as a curved tube, with the base ascending, most of the tube horizontal, and the apex only slightly if at all nodding. In contrast, the flowers of *S. diluvialis* are ascending, and those of *S. graminea* and the closely related *S. nebulorum* are nodding. The lateral sepals of *S. delitescens* curve outward and downward so that the apices are held nearly perpendicular to the axis of the flower. The lateral sepals of the other species are straight for the lower half and may curve outward, inward, or upward. Sheviak also differentiated *S. delitescens* based on pubescence. In *S. delitescens* the trichomes, or hairlike growths, taper toward the apices. In other species, the trichomes are capitate due to the glands at the apex of the trichome being larger in diameter than the stipe. This character, though useful for the study of pressed specimens, is of limited value in the field because of the small size of the hairs.

Distribution

Spiranthes delitescens is the only orchid endemic to Arizona. Because of its proximity to the border with Mexico, it may one day be discovered there also. Currently *S. delitescens* is contained within the extreme Southeast corner of Arizona, where it is known from only Santa Cruz and Cochise Counties. The type specimen of *S. delitescens* is from Santa Cruz County.

Habitat

The elevation gradually rises in the extreme southeastern part of Arizona as the predominantly Sonoran Desert landscape slowly gives way to rolling grasslands, juniper-oak grasslands, and oak woodlands. Permanent streams are likely in the canyons, and occasionally the grasslands are punctuated by desert wetlands called *cienegas*. Cienegas are permanent wetlands, usually maintained by springs or seeps but often with a small stream or creek adding groundwater and surface moisture via runoff. Though cienegas have a wide elevational range, those between 4600 and 5050 feet (1400 and 1530 meters) are of particular interest because some are home to the rare *S. delitescens*. At the time of blooming, the soil near the plants is moist to very wet to boggy. The orchids grow in full sun, but surrounding grasses and sedges protect the foliage. When the plant is in bloom, the flower spikes stick up above the competing growth, but nonblooming plants are nearly impossible to locate. Most of the plants grow on slight to fairly steep slopes, but some also are found on the flats, where they often grow on humps created by cattle in the cienega. On some of the privately owned cienegas harboring *S. delitescens*, cattle have grazed for over 100 years. Though cattle will eat the plants, grazing does not seem to be a long-term detriment, and McClaran and Sundt (1992) speculated it might have a neutral effect or a slight benefit.

Blooming Season

Aboveground growth of *S. delitescens* starts in late spring, and the inflorescence elongates through early summer. The primary blooming season for *S. delitescens* is the month of July. In some years early plants may start to open in late June, and a few flowers may remain fresh into August. A single plant remains in bloom for only about 1 week, but individuals appear over several weeks, resulting in a total blooming season of 4 to 5 weeks. McClaran and Sundt (1992) reported that the life expectancy of *S. delitescens* is only 3 to 4 years after first appearing above ground. They also reported that plants can reappear after an absence of at least 1 year and that the percentage of plants blooming each year varies between 9 and 58 percent. The author and Larry Toolin counted 731 flowering plants at one location in 1999, suggesting that the total number of plants is well over 1000 if McClaran and Sundt's (1992) estimate of nonblooming plants is used.

Conservation

Spiranthes delitescens is federally listed as an endangered species. It is ranked G1/S1 in the state owing to its very few locations. *Spiranthes delitescens* is known from only three locations in Santa Cruz County and one in Cochise County. One colony grows on property owned by the Nature Conservancy; these plants are monitored yearly and have been the subject of studies, including the impact of fire on the population. The other three locations are on private land.

Notes and Comments

What would eventually be called *S. delitescens* was first reported in Arizona by Mason (1971), based on a collection made by Paul S. Martin in O'Donnell Cienaga in 1968. Luer identified it as *Spiranthes graminea*, an orchid previously known only from Mexico, and subsequently published photographs of it in his *Native Orchids of the United States and Canada* (Luer 1975). While working on his study that eventually led to the description of *Spiranthes diluvialis*, Sheviak examined specimens of *S. graminea* from Mexico, along with the supposed *S. graminea* from

Arizona. He decided they represented different taxa and subsequently described *S. delitescens* as a new species (Sheviak 1990).

Following Sheviak's description, it was determined that *S. delitescens* was restricted to Arizona and was extremely rare, with just four known locations. The U.S. Fish and Wildlife Service was petitioned in 1993 to add *S. delitescens* to the endangered species list, and in April 1995 *S. delitescens* was advanced as a candidate for federal recognition as an endangered species. That recognition came in January 1997.

Spiranthes magnicamporum Sheviak
Botanical Museum Leaflets 23: 285. 1973.

Etymology: The name is based on Latin words meaning "large" and "plain," referring to the Great Plains where this species is more common.

Synonymy: none.

Common names: Great Plains ladies'-tresses.

Plate 30

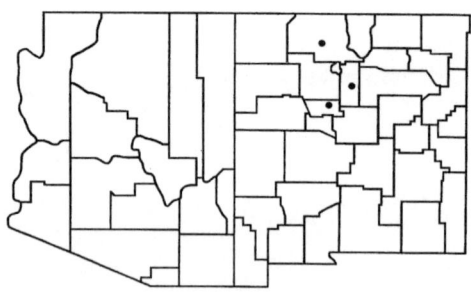

Map 33. Distribution of *Spiranthes magnicamporum*

Description

Plant: 20 to 50 cm tall with three to five basal leaves and multiple bracts above leaves to flowers; stem pubescent near flowers, with 20 to over 50 flowers in dense multiranked spirals.
Roots: three to six, thick, tuberous, 5 cm long × 0.5 cm thick.
Leaves: usually two or three, linear lanceolate to 19 × 1.0 cm; may or may not be present at anthesis.
Floral bracts: ovate, acuminate, 1.0 cm long × 0.5 cm wide.
Flowers: white, tubular, 1.0 cm long × 0.8 cm wide × 0.8 cm high, slightly nodding, strongly scented; petals connivent with dorsal sepal to form hood over column, lateral sepals free and spreading.
Sepals: white, lateral sepals lanceolate, falcate, 2.0 cm long × 0.2 cm wide, finely pubescent on backside; dorsal sepal elliptic lanceolate, 1.0 cm long × 0.25 cm wide.

Petals: white, linear lanceolate to slightly elliptic lanceolate, 1.0 cm long × 0.3 cm wide.

Lip: white with cream to pale yellow center, ovate to oblong, translucent at edges, but thickened and lined with fine papillae in center, two small tubercles at claw with column, apexical third fimbriated.

Column: green, 5 mm long × 2 mm wide, with brown anther cap over two pairs of lanceolate yellow pollinia.

Capsule: plumply ellipsoidal, held erect along the rachis.

Sheviak (1982) included *Spiranthes magnicamporum* (mag-ni-campor'-um) within the *S. cernua* complex, which consists of *S. cernua*, *S. magnicamporum*, *S. ochroleuca*, and *S. odorata*. Identification within that complex is often difficult owing to the extreme variability of *S. cernua* and intergrading due to hybridization. *Spiranthes magnicamporum* is the only member of its complex in either Arizona or New Mexico and is very easy to distinguish from *S. romanzoffiana*, the only other *Spiranthes* species in New Mexico. The flowers of *S. magnicamporum* are much more loosely arrayed, with lateral sepals free and spreading from the base of the flower. Even the apices of the petals are free from the dorsal sepal. On freshly opened flowers the lateral sepals are held out to the side, but as the flower ages, the sepals gradually reach upward, sometimes arching over and sometimes crossing above the hood formed by dorsal sepal and petals. The sepals over the hood are reminiscent of cow's horns. On *S. romanzoffiana* the lateral sepals are connivent with the dorsal sepal and petals to form a tubular hood over and around the column and basal half of the lip. As with *S. romanzoffiana*, the flowers of *S. magnicamporum* are held in three ranks, often in a loose spiral, but sometimes in a dense spiral.

The flowers of *S. magnicamporum* are very strongly scented, giving off an aroma that Luer (1975), Case (1987), and Homoya (1993) compared to coumarin. The lip is another feature that helps distinguish *S. magnicamporum* in the Southwest. The center third of the lip is thick and opaque, in contrast to the rim, which is translucent. This characteristic can be determined easily, even in photographs. Throughout most of its range, and particularly on the Great Plains, the presence or absence of leaves is often used to help identify *S. magnicamporum*. Sheviak (1982) reported that the leaves fade 2 to 3 weeks before the flowers open. The condition of the leaves is not a reliable indicator in New Mexico, however. When the plant is growing in xeric conditions, the leaves fade before blooming, but in moist streamside locations, leaves on nearly half the plants will persist until after flowering. In New Mexico the flowers are visited by bumble bees, which is consistent with Sheviak's (1982) report of pollination by *Bombus fervidus*.

Distribution

Spiranthes magnicamporum is widely distributed in the Great Plains, extending into Texas. It is also prevalent in the Great Lakes region and north into Ontario, Canada. Within those contiguous regions, while not common, it is fairly uniform in distribution. Homoya (1993) also identified four clearly disjunct locations: in Texas, Georgia, Massachusetts, and New Mexico. In New Mexico there is solid evidence placing it in Bernalillo, Santa Fe, and Rio Arriba Counties. The colony in Bernalillo County is the southern limit for the species. The Santa Fe and Rio Arriba occurrences are based on the same population, which grows astride the county line. There are persistent unconfirmed reports of *S. magnicamporum* in Bandelier National Monument in Sandoval County, but that report is not supported with herbarium specimens. Suitable habitat exists in much of north central New Mexico, and *Spiranthes magnicamporum* should be looked for along river and stream systems in much of the area.

Habitat

Through most of its range, *S. magnicamporum* is typically a tall-grass prairie plant, but near the Great Lakes it also grows on exposed sites on bluffs and sand ridges. In New Mexico, it is closely associated with water sources in its two different habitats, at elevations between 5000 and

6000 feet (1524 and 1829 meters). In Bernalillo County it grows in alkaline, crusty soils, at the edge of an abandoned river channel. In Santa Fe and Rio Arriba Counties it grows on saturated sandbars in a river bottom, and higher up above the bank in thick, clay loam. On the sandbars and on the bank it grows among grasses and under shrubs. Sheviak (1991) stated that *S. magnicamporum* is a colonizer of disturbed sites, and the river site is disturbed both by the activities of humans and by the natural shifting of the river channel.

Blooming Season

Spiranthes magnicamporum is one of the last of the orchids in the Southwest to open. It starts blooming in mid to late September and lasts in bloom for about 2 weeks. In other parts of its range, it may remain in bloom until November.

Conservation

Spiranthes magnicamporum is rare in New Mexico and at risk of disappearing. It is on List 1 of rare plants in New Mexico, which means it is considered very rare in the state. Searches by Tom Todsen in 1997 and by the author in 1998 did not turn up any plants at the Bernalillo County location, although William Jennings (2000, personal communication) observed the plants there in 1989. The Santa Fe and Rio Arriba Counties location is healthy with over 50 plants, but the area is subject to abuse by vehicles, and during the 1998 blooming season, heavy equipment was working on the banks and bottom of the river within a few yards of the plants. Attempts should be made to locate other colonies in New Mexico, and if no additional locations are found, positive steps are needed to protect the existing plants from destruction.

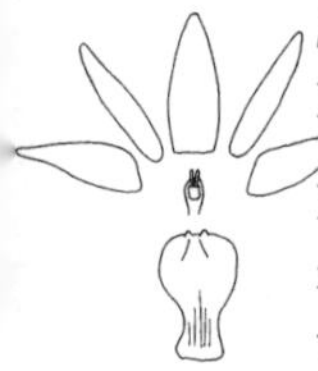

Spiranthes romanzoffiana Chamisso
Linnaea 3: 32. 1828.

Etymology: The specific epithet is in honor of Nicholas Romanzof, a Russian minister of state. The species was discovered during the period when Alaska was Russian territory.

Synonymy:
Gyrostachys romanzowiana (Chamisso) MacMillian, Metaspermae of the Minnesota Valley: 171. 1892.
Orchiastrum romanzoffianum (Chamisso) Greene, Manual of the Botany of the Region of San Francisco Bay: 306. 1894.
Gyrostachys stricta Rydberg, Memoirs New York Botanical Garden 1: 107. 1900.
Ibidium strictum (Rydberg) House, Bulletin Torrey Botanical Club 32: 381. 1905.
Ibidium romanzoffianum (Chamisso) House, Muhlenbergia 1: 129. 1906.
Spiranthes stricta (Rydberg) A. Nelson, New Manual of Botany, Rocky Mountains: 125. 1909.
Triorchis stricta (Rydberg) Nieuwland, American Midland Naturalist 3: 123. 1913.
Triorchis romanzoffiana (Chamisso) Nieuwland, American Midland Naturalist 3: 123. 1913.

Common name: hooded ladies'-tresses.

Plate 31

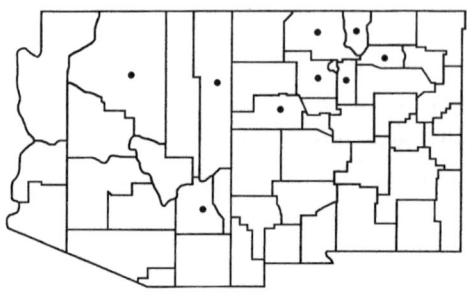

Map 34. Distribution of *Spiranthes romanzoffiana*

Description

Plant: blooming size starts below 10 cm and rarely exceeds 40 cm in the Southwest, but to nearly 60 cm tall elsewhere; leaves mostly basal, but scattered on lower stem on largest plants, with leaflike bracts above; up to 60 flowers in three dense spirals.
Roots: two or three, thick, fibrous, 5 to 10 cm long.
Leaves: three to six, lanceolate; to 18 × 1.1 cm.
Floral bracts: green to whitish, triangular to 1.6 cm long × 0.8 cm wide.
Flowers: white, tubular, with a sweet aroma; sepals and petals form a tight hood over the downward curving spreading lip.
Sepals: lateral sepal white with green suffusion near base, lanceolate, 1.2 × 0.4 cm; dorsal sepal lanceolate, 1 cm long × 0.5 cm wide.
Petals: white, linear, with three faint green stripes, 1.0 cm long × 0.3 cm wide.
Lip: white with five pale green stripes in center, pandurate or fiddle-shaped, with two minor tubercles at base; 1 cm long × 0.6 cm wide at widest part near base.
Column: green, slender, 3 mm long × 1 mm wide, pollinia yellow.
Capsule: ellipsoidal, held in a semierect position.

Spiranthes romanzoffiana (ro-man-zof'-ee-ah-na) appears superficially similar to *S. magnicamporum*, but they can be immediately separated based on flower structure. Both have mostly basal leaves, with the stem of the inflorescence slightly pubescent near the flowers, and both have three rows of flowers in tightly wound spirals. The spirals of *S. romanzoffiana* are neat and tidy because the sepals and petals are tightly connivent to form a hood over the column and basal half of the lip. With *S. magnicamporum* the sepals and petals are free and spreading, resulting in a much fuller-appearing inflorescence. Another way to distinguish these 2 species is by the inclination of the flowers. In *S. romanzoffiana* the flowers are tilted upward at 30 to 45 degrees from the horizontal, while in *S. magnicamporum* the flowers are slightly nodding.

The lip is one of the best features by which to distinguish *S. romanzoffiana*. The lip turns down as it exits the hood formed by the sepals and petals, and the apex curves under for more than 180 degrees, sometimes nearly forming a complete circle. The fiddle-shaped lip narrows noticeably about two-thirds the distance from the base before spreading at the apex. The amount of spreading, and thus the amount of fiddle shape, varies from plant to plant. On some plants the spreading results in the forward part of the lip being as wide as the latter half; in others the lip spreads far less noticeably, to only about one-half the width of the latter half. On older flowers the apex of the lip may dry out and curl up, causing the lip to resemble an anchor more than a fiddle.

An interesting form of *S. romanzoffiana* occurs in Graham County, Arizona. The flowers are identical to the standard form, which also grows in Graham County, but the floral bracts are white or pale whitish green, instead of green. The bracts bend at right angles above the flowers, and from even a short distance, the plants do not appear to be *S. romanzoffiana* because the white bracts suggest a spreading flower. Close inspection of the flowers reveals they are *S. romanzoffiana*, and the different appearance is due to the color of the bracts.

Larson and Larson (1987, 1990) studied the pollination of *S. romanzoffiana*, identifying 11 different pollinators, mostly species of *Bombus* (bumble bees) such as *Bombus bifarius*. One of the attractants for the pollinators is probably the sweet aroma of the flowers. In a pattern similar to that used by pollinators of *Goodyera* species, the bees visit the lowest flowers first and work their way up the inflorescence to recently opened flowers. When the bees depart, they carry pollen from the recently opened flowers to the lowest flowers on the next plant visited, effecting cross-pollination. The seed capsules dehisce in about 6 weeks.

Distribution

Spiranthes romanzoffiana is distributed continuously across Canada and much of the northern United States, particularly New England, the Great Lakes region, and the Northwest. Godfrey (1922) reported a widely disjunct population in Ireland. Along the Pacific Coast of the United States, it extends south into the mountains of southern California. Populations occur along the Rocky Mountains as far south as New Mexico and are in parts of northern Arizona. In Arizona, *S. romanzoffiana* is in the 3 counties of Apache, Coconino, and Graham. The plants in Graham County, the southern limit for the species, appear relatively isolated from others in the state, but their presence suggests *S. romanzoffiana* may eventually be found in Greenlee County. In New Mexico, *S. romanzoffiana* is in Cibola, Mora, Rio Arriba, Sandoval, Santa Fe, and Taos Counties.

Habitat

Of the three *Spiranthes* species in Arizona and New Mexico, *S. romanzoffiana* grows in the highest elevation range. It grows in full sun in habitats at elevations between 7400 and 11,000 feet (2256 and 3353 meters). In the Southwest, *S. romanzoffiana* requires damp to almost wet soil; high mountain meadows are the best place to look for it. Permanently damp spots near the centers of meadows with short grasses may support hundreds of blooming plants, highlighted by the occasional western fringed gentian (*Gentiana thermalis*) and the more common shooting stars (*Dodecatheon pauciflorum*). Nearly as often as it is found in meadows, *S. romanzoffiana* is found along the banks of streams. In these streamside habitats it is likely to be in partial shade from a tree or shrub. Man-made conditions that mimic its natural habitats may also support *S. romanzoffiana*. Moist areas below check dams and along roadside ditches may harbor blooming plants, suggesting an ability to colonize disturbed sites.

Blooming Season

Spiranthes romanzoffiana is a relatively late bloomer. Because its short leaves are well hidden by the grasses and other meadow plants with which it grows, it is very hard to find *S. romanzoffiana* unless it is in bloom. The leaves sprout in midsummer, and spikes are showing by mid to late July. It starts blooming in the first week in August and will continue in bloom until the end of September. By the time *S. romanzoffiana* comes into bloom, *Platanthera* plants that may grow along the same streams will be in fruit. In the forest surrounding the meadows, *Goodyera oblongifolia* and *G. repens* will still be in bloom, and in parts of Arizona and New Mexico, *Malaxis soulei* will be in peak bloom on nearby hillsides.

Conservation

Spiranthes romanzoffiana is well distributed in the Southwest and exists in sufficient numbers to be safe from immediate threats. Its apparent ability to colonize roadside ditches suggests it is adaptable and a willing occupant of disturbed sites. It is not on the list of rare plants for New Mexico, and is ranked G5/S3S4 in Arizona, which means there are between 51 and 100 occurrences over a wide range in the state.

Stenorrhynchos Sprengel

Systema Vegetabilium: 677. 1826.
Etymology: The genus name is derived from Greek words for "narrow" and "snout," referring to the narrow rostellum.

Stenorrhynchos (sten-oh-rin'-kos) occurs from the extreme southern parts of the United States to South America, and in the West Indies. Sprengel established *Stenorrhynchos* to separate from *Spiranthes* those species with larger, showier flowers, a nonspiraled inflorescence, a sharply pointed rostellum, and a mentum formed by the lateral sepals and the base of the column. The plants differ from the related genus *Pelexia* by the structure of the rostellum and the contracted instead of dilated tip of the lip.

Some authors spell the genus name "Stenorrhynchus" but Garay (1982) and Stewart and Griffiths (1995) maintained that the correct spelling is "Stenorrhynchos." Stewart and Griffiths explained that the Greek root word requires a Greek, not a Latin termination. L. C. Richard (1818) first referred to the genus, but he failed to publish a description, merely listing several binomials. He spelled the name "Stenorynchus." When Sprengel properly published the description, he spelled the name "Stenorrhynchos," and referenced Richard as the source of the name.

The subtribe Spiranthinae is large and not completely understood. Many of the genera within the subtribe have been subjected to splitting, recombining, and redefinition by various authors. A history of varied treatment over time is one of the reasons the estimated number of species (between 9 and 60 species) in *Stenorrhynchos* is so imprecise. Lindley

(1830) recognized the genus as distinct from *Spiranthes*. Ames (1924) included *Stenorrhynchos* in *An Enumeration of the Orchids of the United States and Canada*, but later Ames and Schweinfurth (1935) merged it within *Spiranthes*. Correll (1950), Luer (1972, 1975), and Williams (1965) merged *Stenorrhynchos* into an expanded concept of *Spiranthes*, as did the authors of two floristic treatments for Arizona, Kearney and Peebles (1951) and Lehr (1978). Recent scholarship, beginning with Garay (1982), is again recognizing *Stenorrhynchos*. Both Balogh (1982) and Szlachetko (1994) recognized *Stenorrhynchos* but included slightly different species than Garay. For example, Balogh treated *S. michuacanum* as a *Schiedeella*. Balogh's *Schiedeella michuacanum* notwithstanding, if these three recent treatments are followed, only 1 species of *Stenorrhynchos* is native to the United States.

Stenorrhynchos michuacanum (Lexarza) Lindley
Genera and Species of Orchidaceous Plants: 480. 1840.

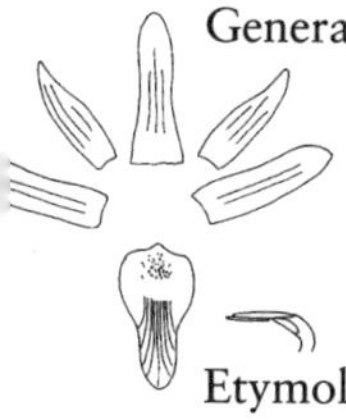

Etymology: The species was named for the Mexican state of Michoacan, in which it was first discovered.

Synonymy:
Neottia michuacana Lexarza, Orchidianum Opusculum: 3. 1825.
Stenorhynchus michuacanus Lindley, Genera and Species of Orchidaceous Plants: 480. 1840.
Spiranthes mechoacana Hemsley, Biologia Centralli-Americana; Botany: 3: 301. 1882–1886.
Schiedeella michuacana Balogh, Orquidea 8(1): 39. 1981.

Common name: none.

Plate 32

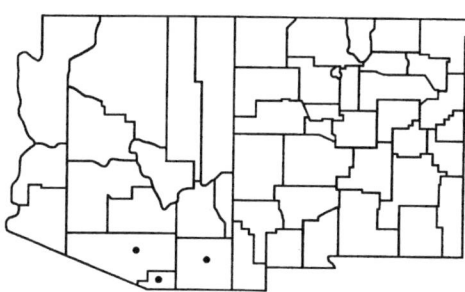

Map 35. Distribution of *Stenorrhynchos michuacanum*

Description

Plant: between 25 and 58 cm tall, with three or four basal leaves becoming bractlike along stem to start of raceme; inflorescence with 10 to 30 tubular flowers.
Roots: few, tuberous, thick.
Leaves: lanceolate to linear lanceolate, to 28 × 3.0 cm.

Floral bracts: lanceolate to triangular, 1.0 cm wide × 2.0 cm tall, sheathing base of ovary, green during bud stage but brown at time of flowering.

Flowers: funnel-shaped, pale green, about 1.5 cm wide × 1.6 cm tall × 2.0 cm long; sepals and petals broadly spreading at their apex.

Sepals: pale green with fine dark green stripes and fine hairs on edges and back; lateral sepals linear, 1.7 cm long × 0.4 cm wide; dorsal sepal linear ovate and slightly concave, 1.6 cm long × 0.5 cm wide.

Petals: pale green with fine dark green stripes, linear falcate, 1.2 cm long × 0.3 cm wide, with fine hairs on edges and back.

Lip: pale green with fine dark green stripes, ovate, tapering to nearly linear above the middle, 1.6 × 0.7 cm, may have yellowish center, rounded at the apex, dilated toward the base, thickens and necks down abruptly to form a claw where it joins the column; lined with fine hairs on deeply V-shaped back third; dilated sides curl up and the apex of the lip curls under 180 degrees.

Column: rounded in the middle, with a long tapering point, 8 mm long × 3 mm wide, four yellow pollinia, long, slender, and wider at one end than the other.

Capsule: ellipsoidal to almost spherical, about 9 × 6 mm, covered with fine hairs.

Stenorrhynchos michuacanum (mich-wa-can'-um) is a stately plant that is easily identifiable even if immature or in early spike. Mature plants have three or four pale green leaves, one of which may be partway up on the stem; non-flowering-size plants have one to three basal leaves. The broad, pale green leaves, slightly reminiscent of either lilies or the Southeast Asian orchid genus *Paphiopedilum*, are unlike those of any other plant in their near-desert habitat. The inflorescence is between 25 and 58 cm tall, and the 10 to 30 flowers face the same direction, generally south to southwest. The flowers are pale green, with dark green stripes on the sepals, petals, and lip. On some plants there is a pale yellow infusion on the inner half of the flower. Fine white hairs line the edges of the sepals and petals and densely cover the backsides of the sepals and the entire ovary.

Correll (1950) reported a strong fragrance of musk and honey, but the plants in Arizona have a very mild musty fragrance noticeable at night. Both the timing of the fragrance and the color of the flowers suggest pollination by night-flying insects. The column structure favors cross-pollination. It is curved so that the pollinator bumps into the stigma at a bend in the column, then while backing out contacts the viscidia and removes pollinia. The seeds are a pale whitish color with dark, almost black embryos.

Distribution

Stenorrhynchos michuacanum is distributed widely in Mexico, but in the United States it occurs only in Texas and Arizona. Its Arizona locations include some fairly close to the border with New Mexico, and it may be found there eventually, but field searches have failed to locate any plants in New Mexico. Within Arizona, it has been reported from only the 3 southeastern counties of Cochise, Pima, and Santa Cruz. One Pima County location, in the Santa Catalina Mountains, is well documented, but *S. michuacanum* has not been seen there since 1949, suggesting it may have been extirpated from that site. However, in September 1999, Nancy Stallcup, a member of the Arizona Native Plant Society, found a new colony in the Santa Catalinas, several miles away from the historical location. About the same time Annita Harlan found a separate colony in the Santa Catalinas, even farther away from the place the plants were discovered in 1949. Plants also grow in other parts of Pima County. J. G. Lemmon collected *S. michuacanum* from the Chiricahua Mountains in Cochise County in 1881. It still grows in at least one canyon there, but Lemmon's annotation on the specimen sheet was not specific to location, so it is not possible to determine if the current colony is the same one he observed. Barneby collected *S. michuacanum* in 1939, supposedly from the Dragoon Mountains in Cochise County, but repeated searches have failed to relocate Barneby's site.

Habitat

Stenorrhynchos michuacanum grows primarily at elevations between 5000 and 6200 feet (1500 and 1900 meters), although its total elevational range extends up to 7000 feet (2134 meters). This region is in transition between the Sonoran desertscrub habitat that is prevalent only about 1000 feet lower in elevation, and the ponderosa pine habitat (montane conifer forest) that becomes common about 7000 feet. Brown (1994) defined this habitat type as Madrean evergreen woodland, consisting of open to dense stands of oak, juniper, and piñon pine. Reminders that this is near-desert habitat abound in the many species of cacti and other Sonoran plants in the vicinity. The most common habitat of *S. michuacanum* is alligator juniper (*Juniperus deppeana* Steud.) woodlands, either in relatively flat terrain or on the sloping sides of drainages and hillsides. In this environment it grows primarily under the junipers, usually just at the drip line, but occasionally also grows under manzanita, oaks, or piñon pines. A few plants grow in the open among grasses within a few yards of junipers. Less often it grows in the bottoms of canyons and drainages, under junipers, oaks, and pines along the banks of seasonal streams. Another habitat is rocky hillsides with a southern exposure, where it grows among or near rocks and just at the drip line of manzanita, oaks, and piñon pines. Other plants in the area include rainbow cactus, cholla, and prickly pear. Still another habitat is grassy oak woodlands where the plants grow among grasses shaded by the oaks. In all of these habitats, the plants may be almost hidden by the surrounding grasses, and when in bloom are difficult to see even from a few feet away. A common associate, essentially an indicator plant, *Milla biflora*, flowers in August and September when *S. michuacanum* is developing, and when found in alligator juniper forest, *M. biflora* is an almost certain sign that the orchid is nearby.

The Pima County location where *S. michuacanum* may have been extirpated was in a slightly different habitat. According to notes made by Lemmon on the specimen sheets, the plants were found in "damp places in pine forests" at an elevation of about 7000 feet (2100 meters).

The collection area is now a public campground. The new locations in the Santa Catalina Mountains discovered by Stallcup and Harlan are at the lower edge of ponderosa pine forest in a mix of scrub oak, pine, and alligator juniper. Most of the orchids are near the junipers, but a few are located under pines with no junipers in sight.

Blooming Season

Aboveground growth starts each year shortly after the onset of the monsoon rains, as early as the first week of July and always by the middle of July. Butterflies feed on nectar from the flowers but have not been observed transporting pollen.

Because of its harsh, dry, habitat, few other orchids grow in association with *S. michuacanum*. In its most common alligator juniper habitat, the only other orchid growing nearby, and that rarely, is *Malaxis soulei*. The pine forest habitat supports a few more orchids such as *Malaxis soulei, Corallorhiza maculata,* and *C. wisteriana*.

Conservation

Stenorrhynchos michuacanum is widely scattered in southeastern Arizona and exists in sufficient numbers to be considered safe from immediate threats. Historically it was thought to be quite rare in Arizona, since it was known from only three locations discovered between 1881 and 1949. The colonies at only one of those original locations may still be extant. About 25 years later a clearer picture of the status of *S. michuacanum* in Arizona began to emerge. Jack Kaiser, a botanist with extensive knowledge of the flora in southeastern Arizona, discovered two locations, with a total of about 40 plants, in the early 1970s. Only Kaiser and a fortunate few members of the Arizona Native Plant Society with whom he shared his knowledge knew those two sites. Then in 1989 Chuck Sheviak discovered leaves of several plants on the south side of the Huachuca Mountains, and in 1993 Mark Fishbein found numerous plants a few kilometers from Sheviak's location. Extensive searches of the Sheviak and Fishbein locations in 1996 turned up nearly 200 plants. Searchers armed with the habitat information from the Kaiser, Sheviak, and Fishbein sites were easily able to find several additional locations nearby. These new locations suggest that the plants found by Sheviak and Fishbein represent portions of a colony several kilometers long that contains many hundreds if not thousands of plants. In 1997, *S. michuacanum* was rediscovered in the Chiricahua Mountains and in 1999 it was rediscovered in the Santa Catalina Mountains. Proper habitat exists in other places in Cochise, Pima, and Santa Cruz Counties, and recent experience suggests that *S. michuacanum* will be found in many more areas.

One possible explanation for the increase in sightings is a gradual change in the floral landscape due to fire suppression. Wildfires that used to periodically sweep the grasslands and foothills were controlled extensively for more than the last hundred years. The fires used to keep the alligator juniper in check, but gradually through the twentieth century the alligator juniper forest spread as the fires were controlled. The increase in alligator juniper woodlands created more suitable habitat for *S. michuacanum* and may be the reason for the apparent increased density of the plants. The U.S. Forest Service has initiated a series of experimental controlled burns in an attempt to mimic the historical fire pattern. Some of the burn areas include groves of alligator juniper, so some habitat for *S. michuacanum* may return to grasslands.

Some of the U.S. Forest Service lands are leased for cattle ranching, and that practice creates a potential threat to the orchids. Cattle eat the orchids along with the grasses, and if cattle are grazing during the growing season for the orchids, the plants usually do not survive to bloom.

Notes and Comments

The secret to finding *S. michuacanum* is to understand its habitat requirements, and then search for it in August and early September, when the leaves are fresh. Many more plants put up leaves than bloom, and the shiny broad leaves are a sharp contrast to the immature narrow grasses that shield them. Anyone waiting to look for the plants when they are in bloom is likely to be disappointed. By then the leaves are faded, the stem has browned, and the inflorescence blends in easily with its now mature and browning herbaceous companions. Even plants observed when in leaf are difficult to pick out amid the clutter of their surroundings, so knowing exactly where to look does not always mean a successful orchid hunt.

Appendix 1. Excluded Species

Arizona

Platanthera dilatata: There is one record of *P. dilatata* from Arizona: a collection made by G. R. Vasey in 1881, specimen number 35076 at the U.S. National Herbarium. Annotations on the specimen sheet say simply "Arizona." Because of uncertainties in Vasey's exact location, and the lack of corroborating data, the potential of *P. dilatata* in Arizona is discounted.

Platanthera hyperborea: All specimens reported to be *P. hyperborea* are actually *P. purpurascens*. All plants in the United States previously understood to be *P. hyperborea* are now known as *P. aquilonis* Sheviak, but even *P. aquilonis* does not occur in Arizona.

Platanthera stricta: All specimens reported to be *P. stricta* are actually *P. purpurascens*. *Platanthera stricta* grows in the northwestern United States and is not known from this area.

New Mexico

Platanthera hyperborea: All plants in the United States previously understood to be *P. hyperborea* are now known as *P. aquilonis* Sheviak.

Platanthera stricta: As in Arizona, all specimens reported to be *P. stricta* are actually *P. purpurascens*. *Platanthera stricta* does not occur in this area.

Spiranthes cernua: This species was listed by Martin and Hutchins (1980), but the plant was most likely *S. magnicamporum*.

Spiranthes vernalis: This species was listed from northeastern New Mexico by Martin and Hutchins' *A Flora of New Mexico, Volume 1* (1980), but according to William Jennings (2000, personal communication), that reference was most likely due to an incorrect interpretation of location, and the specimen was collected much farther east.

Appendix 2. Watch List of Species Not Yet Reported from These States

Arizona

Corallorhiza trifida: Due to its proximity in similar habitats in New Mexico, *C. trifida* may be in the higher elevations of the White Mountains.

Epipactis helleborine: This species is spreading rapidly throughout the United States and Canada. It now has a foothold in New Mexico and should soon be in Arizona.

Piperia unalascensis: This orchid is in Utah and New Mexico and should be sought in northern Arizona.

Platanthera brevifolia: This species grows in New Mexico, not too far from the border with Arizona, and is a potential species for the Chiricahua Mountains.

Platanthera dilatata: This plant should be looked for in northern Arizona in suitable habitats near the Utah and Nevada borders.

Spiranthes diluvialis: This species is in Utah and may have followed waterways into northern Arizona.

New Mexico

Hexalectris revoluta: This plant grows in Texas and Arizona and should be pursued in similar habitats in New Mexico.

Hexalectris warnockii: This species is found in Texas and Arizona, and there is a high expectation that one day *H. warnockii* will be added to the flora of New Mexico, where there also is suitable habitat.

Platanthera obtusata: This plant grows in Colorado within 10 miles (16 km) of New Mexico and should be sought at high elevations in the northern part of the state.

Platanthera zothecina: This orchid occurs in all of the Four Corner states except New Mexico, and since proper habitat exists in New Mexico, it is probably there also.

Stenorrhynchos michuacanum: This species is in Arizona close to the border with New Mexico and is probably in extreme southwestern New Mexico.

Appendix 3. Herbarium Collections Studied

The orchid collections from the following herbaria were used to prepare the distribution, elevation, and blooming season data for this volume. The herbaria were either visited or loaned their collections for study at the University of Arizona.

Herbarium	Symbol
Arizona State University	ASU
Bandelier National Monument	
Desert Museum	DES
Herbarium of Jack Kaiser	
Herbarium of Ronald A. Coleman	
Museum of Northern Arizona	MNA
National Herbarium	US
Navajo Nation	
New Mexico State University	NMS
Northern Arizona University	NAU
Rancho Santa Ana Botanic Garden	RSA
San Juan College	SJC
Sul Ross State University	SRSC
Southwest Research Station	
University of Arizona	ARIZ
University of New Mexico	UNM
University of Texas El Paso	UTEP
U.S. Forest Service, Apache-Sitgreaves National Forest	

Appendix 4. Distribution of Orchids in Arizona

	Apache	Cochise	Coconino	Gila	Graham	Greenlee	La Paz	Maricopa
Calypso bulbosa var. *americana*	X		X			X		
Coeloglossum viride var. *virescens*						X		
Corallorhiza maculata	X	X	X	X	X	X		
Corallorhiza striata	X	X	X	X	X	X		
Corallorhiza wisteriana	X	X	X	X	X	X		
Cypripedium parviflorum var. *pubescens*	X				X	X		
Epipactis gigantea	X	X	X	X	X			X
Goodyera oblongifolia	X	X	X	X	X	X		
Goodyera repens	X					X		
Hexalectris revoluta		X						
Hexalectris spicata var. *arizonica*		X						
Hexalectris spicata var. *spicata*		X						
Hexalectris warnockii		X						
Listera convallarioides	X		X					
Malaxis abieticola		X						
Malaxis corymbosa		X						
Malaxis porphyrea	X	X	X					
Malaxis soulei	X	X	X	X	X	X		
Platanthera limosa		X						
Platanthera purpurascens	X		X		X	X		
Platanthera sparsiflora	X		X	X		X		
Platanthera zothecina	X		X					
Schiedeella arizonica	X	X			X	X		
Spiranthes delitescens		X						
Spiranthes romanzoffiana	X		X		X			
Stenorrhynchos michuacanum		X						

	Mohave	Navajo	Pima	Pinal	Santa Cruz	Yavapai	Yuma
Calypso bulbosa var. *americana*							
Coeloglossum viride var. *virescens*							
Corallorhiza maculata		X	X				
Corallorhiza striata		X	X				
Corallorhiza wisteriana		X	X		X		
Cypripedium parviflorum var. *pubescens*							
Epipactis gigantea	X	X	X		X	X	
Goodyera oblongifolia							
Goodyera repens							
Hexalectris revoluta			X		X		
Hexalectris spicata var. *arizonica*			X		X		
Hexalectris spicata var. *spicata*					X	X	
Hexalectris warnockii							
Listera convallarioides			X				
Malaxis abieticola			X				
Malaxis corymbosa					X		
Malaxis porphyrea			X		X		
Malaxis soulei		X	X		X		
Platanthera limosa			X		X		
Platanthera purpurascens							
Platanthera sparsiflora							
Platanthera zothecina		X					
Schiedeella arizonica			X		X		
Spiranthes delitescens					X		
Spiranthes romanzoffiana							
Stenorrhynchos michuacanum			X		X		

Appendix 5. Distribution of Orchids in New Mexico

	Bernalillo	Catron	Chaves	Cibola	Colfax	Curry	De Baca
Calypso bulbosa var. *americana*	X						
Coeloglossum viride var. *virescens*							
Corallorhiza maculata	X	X		X	X		
Corallorhiza striata	X	X			X		
Corallorhiza trifida							
Corallorhiza wisteriana		X					
Cypripedium parviflorum var. *pubescens*		X			X		
Epipactis gigantea							
Epipactis helleborine	X						
Goodyera oblongifolia	X	X		X			
Goodyera repens		X			X		
Hexalectris nitida							
Hexalectris spicata var. *arizonica*							
Hexalectris spicata var. *spicata*							
Listera cordata							
Malaxis abieticola		X					
Malaxis porphyrea		X			X		
Malaxis soulei		X					
Piperia unalascensis							
Platanthera aquilonis							
Platanthera brevifolia		X					
Platanthera huronensis							
Platanthera limosa		X					
Platanthera purpurascens		X			X		
Platanthera sparsiflora							
Schiedeella arizonica		X					
Spiranthes magnicamporum	X						
Spiranthes romanzoffiana			X				

	Dona Ana	Eddy	Grant	Guadalupe	Harding	Hidalgo	Lea
Calypso bulbosa var. *americana*							
Coeloglossum viride var. *virescens*			X				
Corallorhiza maculata			X				
Corallorhiza striata			X				
Corallorhiza trifida							
Corallorhiza wisteriana			X				
Cypripedium parviflorum var. *pubescens*			X				
Epipactis gigantea		X	X		X		
Epipactis helleborine							
Goodyera oblongifolia							
Goodyera repens							
Hexalectris nitida		X					
Hexalectris spicata var. *arizonica*							
Hexalectris spicata var. *spicata*							
Listera cordata							
Malaxis abieticola			X				
Malaxis porphyrea			X				
Malaxis soulei			X				
Piperia unalascensis							
Platanthera aquilonis							
Platanthera brevifolia			X				
Platanthera huronensis							
Platanthera limosa							
Platanthera purpurascens							
Platanthera sparsiflora			X				
Schiedeella arizonica			X				
Spiranthes magnicamporum							
Spiranthes romanzoffiana							

	Lincoln	Los Alamos	Luna	McKinley	Mora	Otero	Quay
Calypso bulbosa var. *americana*	X				X		
Coeloglossum viride var. *virescens*							
Corallorhiza maculata	X	X			X	X	
Corallorhiza striata	X	X				X	
Corallorhiza trifida							
Corallorhiza wisteriana	X					X	
Cypripedium parviflorum var. *pubescens*		X				X	
Epipactis gigantea		X				X	
Epipactis helleborine							
Goodyera oblongifolia		X			X	X	
Goodyera repens		X					
Hexalectris nitida							
Hexalectris spicata var. *arizonica*						X	
Hexalectris spicata var. *spicata*							
Listera cordata					X		
Malaxis abieticola						X	
Malaxis porphyrea	X					X	
Malaxis soulei		X					
Piperia unalascensis				X			
Platanthera aquilonis	X						
Platanthera brevifolia	X					X	
Platanthera huronensis							
Platanthera limosa							
Platanthera purpurascens	X						
Platanthera sparsiflora	X					X	
Schiedeella arizonica	X					X	
Spiranthes magnicamporum							
Spiranthes romanzoffiana					X		

	Rio Arriba	Roosevelt	San Juan	San Miguel	Sandoval	Santa Fe
Calypso bulbosa var. *americana*	X			X	X	X
Coeloglossum viride var. *virescens*				X		
Corallorhiza maculata	X		X	X	X	X
Corallorhiza striata	X			X	X	X
Corallorhiza trifida	X			X		
Corallorhiza wisteriana				X		X
Cypripedium parviflorum var. *pubescens*			X	X		X
Epipactis gigantea			X		X	
Epipactis helleborine						
Goodyera oblongifolia	X			X	X	X
Goodyera repens				X	X	
Hexalectris nitida						
Hexalectris spicata var. *arizonica*						
Hexalectris spicata var. *spicata*						
Listera cordata	X			X		
Malaxis abieticola						
Malaxis porphyrea					X	
Malaxis soulei					X	
Piperia unalascensis						
Platanthera aquilonis	X			X	X	X
Platanthera brevifolia						
Platanthera huronensis	X			X		
Platanthera limosa						
Platanthera purpurascens	X			X	X	
Platanthera sparsiflora					X	
Schiedeella arizonica						
Spiranthes magnicamporum	X					X
Spiranthes romanzoffiana	X				X	X

	Sierra	Socorro	Taos	Torrance	Union	Valencia
Calypso bulbosa var. *americana*		X	X	X		X
Coeloglossum viride var. *virescens*	X		X			
Corallorhiza maculata	X	X	X	X	X	X
Corallorhiza striata			X	X		
Corallorhiza trifida			X			
Corallorhiza wisteriana						
Cypripedium parviflorum var. *pubescens*						
Epipactis gigantea						
Epipactis helleborine						
Goodyera oblongifolia			X			
Goodyera repens	X		X			
Hexalectris nitida						
Hexalectris spicata var. *arizonica*						
Hexalectris spicata var. *spicata*	X					
Listera cordata			X			
Malaxis abieticola						
Malaxis porphyrea	X					
Malaxis soulei	X	X				X
Piperia unalascensis						
Platanthera aquilonis						
Platanthera brevifolia	X					
Platanthera huronensis			X			
Platanthera limosa						
Platanthera purpurascens			X			
Platanthera sparsiflora						
Schiedeella arizonica	X					
Spiranthes magnicamporum						
Spiranthes romanzoffiana			X			

Appendix 6. The Counties of Arizona and New Mexico

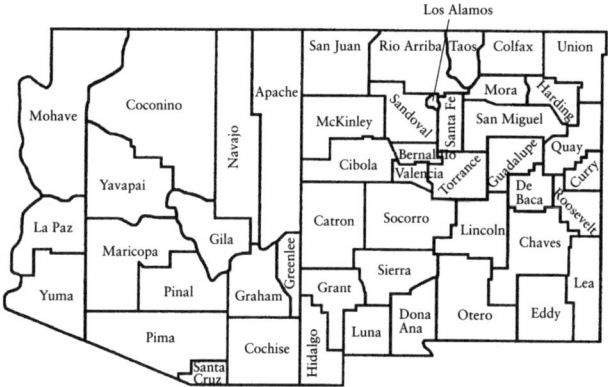

Map 36. The Counties of Arizona and New Mexico

Bibliography

Abrams, L. 1940. *Illustrated Flora of the Pacific States*. Stanford University Press, Stanford, Calif.

Ackerman, J. D. 1975. Reproductive Biology of *Goodyera oblongifolia* (Orchidaceae). *Madroño* 23 (4): 191–198.

———. 1977. Biosystematics of the Genus *Piperia* Rydb. *Botanical Journal of the Linnean Society* 75: 245–270.

Ackerman, J. D., and M. R. Mesler. 1979. Pollination Biology of *Listera cordata* (Orchidaceae). *American Journal of Botany* 66 (7): 820–824.

Ames, O. 1910. *Orchidaceae*, Vol. IV: *The Genus Habenaria in North America*. Merrymount Press, Boston [reprinted in 1979].

———. 1924. *An Enumeration of the Orchids of the United States and Canada*. American Orchid Society, Boston.

Ames, O., and D. Correll. 1943. Notes on North American Orchids. *Botanical Museum Leaflets*, Harvard University 11 (1): 1–2.

Ames, O., and C. Schweinfurth. 1935. Nomenclatural Studies in Malaxis and Spiranthes. *Botanical Museum Leaflets*, Harvard University 3 (8): 128.

Arditti, J. 1992. *Fundamentals of Orchid Biology*. John Wiley and Sons, New York.

Arizona Game and Fish Department. 1999. Status Designations. Unpublished abstract compiled and edited by the Heritage Data Management System, Arizona Game and Fish Department, Phoenix.

Ashmore, S. 1995. Origins of *Calypso bulbosa* Aaron Island. *North American Native Orchid Journal* 1 (2): 87–92.

Atwood, J. T. 1984. The Relationships of the Slipper Orchids (Subfamily Cypripedioideae). *Selbyana* 7 (2, 3, 4): 129–247.

Baldwin, H. 1884. *The Orchids of New England*. John Wiley and Sons, New York.

Balogh, P. 1981. Nomenclature Notes on the Genus *Schiedeella* Schlechter (Orchidaceae) *Orquidea* 8: 38–40.

———. 1982. Generic Redefinition in Subtribe Spiranthinae (Orchidaceae). *American Journal of Botany* 69 (7): 1119–1132.

Bennett, P. S., R. R. Johnson, and M. R. Kunzmann. 1996. *An Annotated List of Vascular Plants of the Chiricahua Mountains*. U.S. Geological Survey, University of Arizona, Tucson.

Bingham, M. T. 1939. *Orchids of Michigan*. Cranbrook Institute of Science, Bloomfield Hills, Mich.

Boyden, T. C. 1982. The Pollination Biology of *Calypso bulbosa* var. *americana* (Orchidaceae): Initial Deception of Bumblebee Visitors. *Oecologia* 55: 175–184.

Brackley, F. E. 1985. The Orchids of New Hampshire. *Rhodora* 87: 849.

Brown, D. E., ed. 1994. *Biotic Communities Southwestern United States and Northwestern Mexico*. University of Utah Press, Salt Lake City.

Brunton, D. F. 1986. Status of the Giant Helleborine, *Epipactis gigantea* (Orchidaceae), in Canada. *Canadian Field-Naturalist* 100 (3): 414–417.

Burns-Balough, P. 1989. *Schiedeella dodii* Burns-Balogh—New Species from the Dominican Republic. *Die Orchidee* 40 (5): 171–173.

Calder, J. A., and R. L. Taylor. 1968. *Flora of the Queen Charlotte Islands*. Research Branch, Canadian Department of Agriculture, Ottawa, Ontario, 287–300.

Case, F. W. 1964. *Orchids of the Western Great Lakes Region*. Cranbrook Institute of Science, Bloomfield Hills, Mich.

———. 1987. *Orchids of the Western Great Lakes Region*, revised edition. Cranbrook Institute of Science, Bloomfield Hills, Mich.

Catling, P. M. 1983. Autogamy in Eastern Canadian Orchidaceae. *Le naturaliste Canadien* 110: 37–53.

Catling, P. M., and V. Catling. 1989. Observations of the Pollination of *Platanthera huronensis* in Southwest Colorado. *Lindleyana* 4 (2): 78–84.

———. 1991. Synopsis of Breeding Systems and Pollination in North American Orchids. *Lindleyana* 6 (4): 187–210.

———. 1997. Morphological Discrimination of *Platanthera huronensis* in the Canadian Rocky Mountains. *Lindleyana* 12 (2): 72–78.

Catling, P. M., and V. S. Engel. 1993. Systematics and Distribution of *Hexalectris spicata* var. *arizonica* (Orchidaceae). *Lindleyana* 8 (3): 119–125.

Chapman, W. K. 1997. *Orchids of the Northeast, a Field Guide*. Syracuse University Press, Syracuse, N.Y.

Coleman, R. A. 1995. *The Wild Orchids of California*. Cornell University Press, Ithaca, N.Y.

———. 1999. *Hexalectris revoluta* in Arizona. *North American Native Orchid Journal* 5 (4): 312–315.

Correll, D. S. 1943. The Genus *Habenaria* in Western North America. *Leaflets of Western Botany* 3: 233–256.

———. 1950. *Native Orchids of North America*. Chronica Botanica Company, Waltham, Mass.

Coulter, J. M. 1909. *New Manual of Botany of the Rocky Mountains*. American Book Company, New York.

Cribb, P. 1997. *The Genus Cypripedium*. Timber Press, Portland, Ore.

Davies, P. H., J. A. Davies, and A. Huxley. 1988. *Wild Orchids of Britain and Europe*. Hogarth Press, London.

Dod, D. 1978. *Spiranthes fauci-sanguinea* Dod. *Moscosoa* 1 (3): 60–62.

Dressler, R. L. 1981. *The Orchids, Natural History and Classification*. Harvard University Press, Cambridge, Mass.

———. 1993. *Phylogeny and Classification of the Orchid Family*. Dioscorides Press, Portland, Ore.

Epple, A. O. 1995. *A Field Guide to the Plants of Arizona*. Falcon Press Publishing Company, Helena, Mont.

Farwell, O. A. 1923. Notes on the Michigan Flora. In: P. Welch and E. McCartney, eds., *Papers of the Michigan Academy of Science* 1: 92.

Fernald, M. L. 1946. Identification of North American Plants. *Rhodora* 48: 193–197.

Freudenstein, J. V. 1992. Systematics of *Corallorhiza* and the Corallorhizinae (Orchidaceae). Ph.D. thesis, Cornell University [unpublished].

———. 1996. Proposal to Conserve the Name *Corallorhiza* Gagnebin (Orchidaceae) with a Conserved Spelling. *Taxon* 45: 695–696.

———. 1997. A Monograph of *Corallorhiza* (Orchidaceae). In: G. A. Romero and O. H. Pfister, eds., *Harvard Papers in Botany* 10: 5–52.

Garay, L. S. 1982. A Generic Revision of the *Spiranthinae*. *Botanical Museum Leaflets, Harvard University* 28 (4): 278–425.

Gibson, W. H. 1905. *Our Native Orchids*. Doubleday, Page, and Company, New York.

Godfrey, M. J. 1922. *Spiranthes romanzoffiana*. *Orchid Review* 30: 261–264.

Greenman, J. M. 1903. New and Otherwise Noteworthy Angiosperms from Mexico and Central America. *Proceedings of the American Academy of Arts and Sciences* 39 (5): 77.

Greenwood, E. W. 1981. Mexican Terrestrial Orchids. In: E. H. Plaxton, ed., *North American Terrestrial Orchids*. Michigan Orchid Society, Southfield, Mich., 81–86.

Grier, R. 1984. Wild Orchids of the UK — The Frog Orchid (*Coeloglossum viride*). *Orchid Review* 92 (1090): 248–249.

Hawkes, A. D. 1965. *Encyclopaedia of Cultivated Orchids*. Faber and Faber, London.

Hitchcock, C. L. 1969. *Vascular Plants of the Pacific Northwest*. University of Washington Press, Seattle.

Homoya, M. A. 1993. *Orchids of Indiana*. Indiana Academy of Science, Bloomington.

Judd, W. W. 1971. Wasps Pollinating Helleborine, *Epipactis helleborine* (L.) Crantz, at Owens Sound, Ontario. *Proceedings, Entomological Society of Ontario* 102: 115–118.

Kearney, T. H., and R. Peebles. 1951. *Arizona Flora*. University of California Press, Berkeley.

Keenan, P. E. 1998. *Wild Orchids across North America*. Timber Press, Portland, Ore.

Larson, K. S., and R. J. Larson. 1990. Lure of the Locks: Showiest Ladies'-tresses Orchids, *Spiranthes romanzoffiana*, Affect Bumblebee, *Bombus* ssp., Foraging Behavior. *Canadian Field-Naturalist* 104 (4): 519–525.

Larson, R. J., and K. S. Larson. 1987. Observations on the Pollination Biology of *Spiranthes romanzoffiana*. *Lindleyana* 2 (4): 176–179.

Lehr, J. H. 1978. *A Catalogue of the Flora of Arizona*. Desert Botanical Garden, Phoenix.

Light, M., and M. MacConaill. 1989. Albinism in *Platanthera hyperborea*. *Lindleyana* 4 (3): 158–160.

———. 1990. Population Dynamics of a Terrestrial Orchid. In: *Proceedings of the 13th World Orchid Conference*. 1990 World Orchid Conference Trust, Auckland, N.Z., 245–247.

———. 1991. Patterns of Appearance in *Epipactis helleborine* (L.) Crantz. In: T. C. E. Wells and J. H. Willems, eds., *Population Ecology of Terrestrial Orchids*. SPB Publishing, The Hague, 77–87.

Lindley, J. 1830. *The Genera and Species of Orchidaceous Plants*. Ridgeway, London.

Long, R. 1980. *Calypso bulbosa* (Linnaeus) Oakes in Z. Thompson. *Davidsonia* 11 (1): 13–16.

Luer, C. 1972. *The Native Orchids of Florida*. New York Botanical Garden, New York.

Luer, C. A. 1975. *The Native Orchids of the United States and Canada*. New York Botanical Garden, New York.

Martin, W. C., and C. R. Hutchins. 1980. *A Flora of New Mexico*. J. Cramer, Vaduz, Germany.

Mason, C. T. 1971. Notes on the Flora of Arizona. *Journal of the Arizona Academy of Science* 6 (3): 189.

McClaran, M. P., and P. C. Sundt. 1992. Population Dynamics of the Rare Orchid, *Spiranthes delitescens*. *Southwestern Naturalist* 37 (3): 299–303.

McVaugh, R. 1985. *Flora Novo-Galiciana*, Vol. 16: *Orchidaceae*. University of Michigan Press, Ann Arbor.

Morgan, R., and J. Ackerman. 1990. Two New Piperias (Orchidaceae) from Western North America. *Lindleyana* 5 (4): 205–211.

Morgan, R., and L. Glicenstein. 1993. Additional California Taxa in *Piperia* (Orchidaceae). *Lindleyana* 8 (2): 89–95.

Morris, F., and A. E. Eames. 1929. *Our Wild Orchids*. Charles Scribner Sons, New York.

Mosquin, T. 1970. The Reproductive Biology of *Calypso bulbosa* (Orchidaceae). *Canadian Field-Naturalist* 84: 291–296.

Mousley, H. 1924. Calypso. *Journal of the New York Botanical Garden* 25: 25–31.

———. 1927. The Genus *Amesia* in North America. *Canadian Field-Naturalist* 41 (1): 28–31.

Nieuwland, J. A. 1913. An Older Name for *Listera*. *American Midland Naturalist* 3: 128–129.

Niles, G. G. 1904. *Bog Trotting for Orchids*. G. P. Putman's Sons, New York.

Pinkava, D. J., E. Lehto, T. Reeves, and E. Sundell. 1975. Plants New to Arizona. *Journal of the Arizona Academy of Science* 10 (3): 146–147.

Proctor, H. C., and L. D. Harder. 1995. Effect of Pollination Success on Floral Longevity in the Orchid *Calypso bulbosa* (Orchidaceae). *American Journal of Botany* 82 (9): 1131–1136.

Ramsey, C. T. 1950. The Triggered Rostellum of the Genus *Listera*. *American Orchid Society Bulletin* 19: 482–485.

Rasmussen, H. N. 1995. *Terrestrial Orchids, from Seed to Mycotrophic Plant*. Cambridge University Press, New York.

Reddoch, J. M., and A. H. Reddoch. 1997. The Orchids of the Ottawa District: Floristics, Phytogeography, Population Studies, and Historical Review. *Canadian Field-Naturalist* 111 (1): 1–185.

Richard, L. C. 1818. De Orchideis Europaeis. *Mémoires Du Muséum D'Histoire Naturelle* 4: 59.

Ross, E. S. 1988. Does *Epipactis gigantea* Mimic Aphids? *Fremontia* 16 (2): 28–29.

Rydberg, P. A. 1901. The American Species of *Limnorchis* and *Piperia*, North of Mexico. *Bulletin of the Torrey Botanical Club* 28: 605–643.

Salazar, G. A. 1991. *Hexalectris warnockii* (Orchidaceae): Primer Registro Para Mexico. *Acta Botanica Mexicana* 16: 1–5.

———. 1993. *Malaxis wendtii*, a New Orchid Species from Coahuila and Nuevo Leon, Mexico. *Orquidea* (Mexico) 13 (1–2): 281–284.

Salazar, G. A., and S. Arenas. 2001. Nomenclatural Changes in Mexican Orchidaceae. *Lindleyana* 16 (3): 149–150.

Schrenk, W. J. 1978. North American Platantheras: Evolution in the Making. *American Orchid Society Bulletin* 47 (3): 429–437.

Sheehan, J. 1992. *Goodyera* Propagation. *American Orchid Society Bulletin* 61 (9): 892.

Sheviak, C. J. 1982. *Biosystematic Study of the* Spiranthes cernua *Complex*. New York State Museum, Bulletin 448, Albany.

———. 1990. A New *Spiranthes* (Orchidaceae) from the Cienegas of Southernmost Arizona. *Rhodora* 92 (872): 213–229.

———. 1991. Morphological Variation in the Compilospecies *Spiranthes cernua* (L) L. C. Rich.: Ecologically Limited Effects of Gene Flow. *Lindleyana* 6 (4): 228–234.

———. 1993. *Cypripedium parviflorum* Salisb. var. *makasin* (Farwell) Sheviak. *American Orchid Society Bulletin* 62 (4): 403.

———. 1994. *Cypripedium parviflorum* Salisb. I: The Small-flowered Varieties. *American Orchid Society Bulletin* 63 (6): 664–669.

———. 1995. *Cypripedium parviflorum* Salisb. Part II: The Large Flowered Plants and Patterns of Variation. *American Orchid Society Bulletin* 64 (6): 606–612.

———. 1999a. *Platanthera hyperborea* and a Reappraisal of the Green *Platantheras*. *North American Native Orchid Journal* 5 (2): 117–141.

———. 1999b. The Identities of *Platanthera hyperborea* and *P. huronensis*, with the Description of a New Species from North America. *Lindleyana* 14 (4): 193–203.

———. 2000. Refinements in Our Understanding of Some Green Platantheras. *North American Native Orchid Journal* 6 (2): 88–92.

Sheviak, C. J., and W. Jennings. 1997. *Platanthera purpurascens*. *North American Native Orchid Journal* 3 (4): 445–449.

Sivinski, R., and K. Lightfoot, eds. 1995. *Inventory of the Rare and Endangered Plants of New Mexico*. New Mexico Forestry and Resources Conservation Division Energy, Minerals and Natural Resources Department, Santa Fe, New Mexico.

Smith, W. R. 1993. *Orchids of Minnesota*. University of Minnesota Press, Minneapolis.

Steele, W. K. 1996. Large Scale Seedling Production of North American *Cypripedium* Species. In: C. Allen, ed., *North American Native Terrestrial Orchids, Propagation and Production*. North American Native Terrestrial Orchid Conference, Germantown, Maryland.

Stewart, J. 1996. *Orchids of Kenya*. Timber Press, Portland, Ore.

Stewart, J., and M. Griffiths. 1995. *Manual of Orchids*. Timber Press, Portland, Ore.

Soto, A. 1988. Listado de las orquideas de Mexico. *Orquidea* (Mexico) 11: 254.

Summerhayes, V. S. 1968. *Wild Orchids of Britain*. Collins, London.

Summers, B. 1981. *Missouri Orchids*. Missouri Department of Conservation.

Szczawinski, A. F. 1975. *The Orchids of British Columbia*. British Columbia Provincial Museum, Victoria, Canada.

Szlachetko, D. L. 1992. Genera and Species of the Subtribe Spiranthinae (Orchidaceae). 2. A Revision of *Schiedeella*. *Fragmenta Floristica et Geobotanica* 37 (1): 157–204.

———. 1993. *Schiedeella romeroana* (Orchidaceae, Spiranthinae), a New and Interesting Species from Mexico. *Rhodora* 95: 1–5.

———. 1994. Studies on the Spiranthinae (Orchidaceae). I. Miscellanea. *Fragmenta Floristica et Geobotanica* 39 (2): 417–438.

Taylor, D. L., and T. D. Bruns. 1997. Independent, Specialized Invasions of Ectomycorrhizal Mutualism by Two Nonphotosynthetic Orchids. *Proceedings of the National Academy of Sciences of the United States of America* 95: 4510–4515.

Tidestrom, I., and T. Kittell. 1941. *Flora of Arizona and New Mexico*. Catholic University of America Press, Washington, D.C.

Todsen, T. A., and T. K. Todsen. 1971. Color Variation of *Corallorhiza* in New Mexico. *Southwestern Naturalist* 16 (1): 121–122.

Todsen, T. K. 1995. *Malaxis wendtii* (Orchidaceae) in the United States. *Sida* 16 (3): 591.

———. 1997. Naming a Southwestern *Malaxis* (Orchidaceae). *North American Native Orchid Journal* 3 (3): 335–339.

van der Pijl, L., and C. Dodson. 1966. *Orchid Flowers, Their Pollination and Evolution*. University of Miami Press, Coral Gables, Fla.

Weber, W. A. 1989. New Names and Combinations, Principally in the Rocky Mountain Flora—VII. *Phytologia* 67 (6): 425–428.

Wentworth, T. K. 1982. Vegetation and Flora of the Mule Mountains, Cochise County, Arizona. *Journal of the Arizona-Nevada Academy of Science* 17: 29–44.

Whiting, R. E., and P. M. Catling. 1986. *Orchids of Ontario*. Canacoll Foundation, Ottawa, Canada.

Williams, L. O. 1934. Field and Herbarium Studies III. *Annals of the Missouri Botanical Garden* 21: 343.

———. 1937. The Orchidaceae of the Rocky Mountains. *American Midland Naturalist* 18 (5): 830–841.

———. 1951. Orchidaceae of Mexico. *Ceiba* 2 (1–4): 1–274.

———. 1965. *The Orchidaceae of Mexico*. Escuela Agricola Panamericana, Tegucigalpa, Honduras.

Zelmer, C. D., and R. S. Currah. 1995. Evidence for a Fungal Liaison between *Corallorhiza trifida* (Orchidaceae) and *Pinus contorta* (Pinaceae). *Canadian Journal of Botany* 73: 862–866.

Index of Scientific and Common Names

Page numbers in bold face indicate the main text descriptions.

Achroanthes
 corymbosa, 128
 montana, 136
Aconitum coclumbianum, 179
Alaska piperia. See *Piperia unalascensis*
alcove bog-orchid. See *Platanthera zothecina*
Amesia
 gigantea, 70
 latifolia, 75
Aphrodite's foot. See *Cypripedium*
Aplectrum, 19
Aquilegia chrysantha, 179
Arethusa spicata, 103

Bifolium, 113
 convallarioides, 114
 cordatum, 118
Bombus, 23, 206
 bifarius, 206
 fervidus, 202
 occidentalis, 82
bracted green orchis. See *Coeloglossum viride* var. *virescens*
broad-leaved helleborine. See *Epipactis helleborine*
broad-leaved twayblade. See *Listera convallarioides*

calypso. See *Calypso bulbosa*
Calypso, 17, **19**
 americana, 21
 borealis, 21
 bulbosa, 3, 5, 7, 8, 10, 13, 15, 19, 21, 32, 51, 84, 88, 121, 139, 161
 bulbosa var. *americana*, 11, 12, 19, **21–26**, 225, 227, plate 1
 bulbosa var. *americana* f. *candida*, 23
 bulbosa var. *americana* f. *rosea*, 23
 bulbosa var. *bulbosa*, 19
 bulbosa var. *occidentalis*, 19, 26
 bulbosa var. *speciosa*, 19
Calypsoeae, 19
Canelo Hills ladies' tresses. See *Spiranthes delitescens*
Carnegiea gigantea, 73
chatterbox. See *Epipactis gigantea*
Cladorhiza maculata, 37
Coccygonimus pedalis, 45
cock's comb. See *Hexalectris spicata*
Coeloglossum, 17, **27**, 147
 bracteatum, 28
 viride, 3, 5, 8, 15, 25, 51, 66, 121
 viride var. *bracteatum*, 28
 viride var. *virescens*, 11, 12, **28–33**, 225, 227, plate 2
 viride var. *viride*, 29, 31
Corallorhiza, 7, 8, 10, 17, 19, **35–36**, 84, 91, 126, 162
 arizonica, 103
 bigelovii, 43
 corallorhiza, 48
 corallorhiza ssp. *coloradensis*, 48
 grabhamii, 37
 innata, 48
 innata var. *virescens*, 48

Corallorhiza (continued)
  involuta, 45
  macraei, 43
  maculata, 5, 6, 7, 8, 9, 10, 11, 12, 25, 32, 36, **37–42**, 46, 49, 51, 53, 54, 55, 56, 57, 66, 74, 117, 121, 126, 134, 139, 146, 156, 167, 171, 175, 189, 215, 225, 227, plates 3 and 4
  maculata var. *flavida*, 39
  maculata var. *fusca*, 39, 40
  maculata var. *immaculata*, 39
  maculata var. *intermedia*, 39, 40
  maculata var. *maculata*, 39
  maculata var. *mexicana*, 39, 41, plate 3
  maculata var. *occidentalis*, 39, 41
  maculata var. *punicia*, 39, 40
  mexicana, 37
  multiflora, 37
  ochroleuca, 43
  odontorhiza, 53
  spicata, 103
  striata, 5, 7, 8, 10, 11, 12, 25, 32, 36, 41, **43–47**, 51, 53, 54, 56, 57, 104, 105, 117, 126, 134, 139, 156, 167, 171, 175, 189, 225, 227, plates 4 and 5
  striata forma *fulva*, 45
  striata var. *flavida*, 45
  striata var. *involuta*, 45
  striata var. *striata*, 44
  striata var. *vreelandii*, 43, 45, 47
  trifida, 3, 5, 6, 10, 11, 12, 16, 36, 41, **48–52**, 54, 221, 227, plate 5
  trifida var. *verna*, 48, 51
  unguiculata, 53
  verna, 48, 50
  vreelandii, 43, 45
  wisteriana, 3, 5, 8, 10, 11, 12, 25, 32, 36, 38, 41, 46, 49, 51, **53–57**, 67, 74, 107, 117, 126, 134, 139, 167, 171, 187, 189, 215, 225, 227, plates 6 and 7
  wyomingensis, 48
Corallrrhiza, 36
coralroot. See *Corallorhiza*
Correll's cock's comb. See *Hexalectris revoluta*
creeping goodyera. See *Goodyera repens*
creeping ladies' tresses. See *Goodyera repens*
crested coralroot. See *Hexalectris spicata*
curly coralroot. See *Hexalectris revoluta*

Cypripedioideae, 59
Cypripedium, 18, 19, 22, **59–60**
  bulbosum, 21
  calceolus, 61, 64
  calceolus var. *planipetalum*, 61
  calceolus var. *pubescens*, 61, 64
  flavescens, 61
  hirsutum, 61, 64
  parviflorum, 3, 5, 8, 9, 10, 32, 41, 51, 74, 84, 161, 171
  parviflorum var. *makasin*, 64
  parviflorum var. *parviflorum*, 64
  parviflorum var. *pubescens*, 11, 12, 15, 16, 60, **61–68**, 225, 227, plate 8
  pubescens, 16, 61, 64
  veganum, 61, 64
Cytherea
  borealis, 21
  bulbosa, 21

deer's head orchid. See *Calypso bulbosa* var. *americana*
Dodecatheop
  pauciflorum, 207
  pulchellum, 66
draft rattlesnake plantain. See *Goodyera repens*
dragon's claw. See *Hexalectris spicata*

early coralroot. See *Corallorhiza trifida*
Epipactis, 17, 69
  americana, 70
  convallarioides, 114, 118
  decipiens, 80
  gigantea, 4, 7, 9, 11, 12, 13, 16, 69, **70–74**, 76, 96, 107, 180, 225, 227, plate 9
  helleborine, 5, 10, 11, 12, 13, 69, **75–78**, 221, 227, plate 9
  helleborine forma *monotropoides*, 76
  helleborine var. *viridens*, 75
  latifolia, 75

fairy slipper. See *Calypso bulbosa* var. *americana*
false lady's slipper. See *Epipactis gigantea*
frog orchid. See *Coeloglossum viride* var. *virescens*

Gentiana
  grandis, 166
  thermalis, 207

giant helleborine. See *Epipactis gigantea*
Glass Mountain coralroot. See *Hexalectris nitida*
Goodyera, 7, 10, 17, 41, **79**, 206
  decipiens, 80
  menziesii, 80
  oblongifolia, 5, 8, 11, 12, 26, 32, 46, 51, 57, 67, 74, 79, **80–84**, 86, 87, 88, 117, 121, 134, 139, 161, 171, 176, 189, 208, 225, 227, plate 10
  oblongifolia var. *reticulata*, 81
  ophilides, 85
  repens, 3, 5, 8, 11, 12, 15, 26, 32, 46, 67, 79, 83, **85–89**, 117, 121, 161, 171, 208, 225, 227, plate 11
  repens var. *ophilides*, 85, 87
  tesselata, 87
Goodyerinae, 79
Great Plains ladies' tresses. See *Spiranthes magnicamporum*
green bog-orchid. See *Platanthera aquilonis*
Gyrostachys
  romanzowiana, 204
  stricta, 204

Habenaria, 27, 141, 147, 157, 171
  aggregata, 173
  bracteata, 28
  brevifolia, 154
  huronensis, 158
  hyperborea, 151, 152, 153, 162, 172
  hyperborea var. *huronensis*, 158
  hyperborea var. *purpurascens*, 168
  leucostachys var. *viridis*, 173
  limosa, 164
  purpurascens, 172
  saccata, 172
  schischmareffiana, 143
  sparsiflora, 173
  sparsiflora var. *brevifolia*, 154
  thurberi, 164
  unalaschcensis, 143
  viridis, 28
  virescens, 28
  zothecina, 177
Habenariinae, 27
heart-leaved twayblade. See *Listera cordata*
Helleborine gigantea, 70
Herminium unalascensis, 143
Hexalectris, 5, 8, 17, 36, **91**
  grandiflora, 91
  mexicana, 91
  nitida, xiii, 3, 10, 11, 12, 13, 16, 78, 92, **93–97**, 227, plate 12
  parviflora, 91
  revoluta, 3, 11, 12, 15, 92, **98–102**, 107, 221, 225, plates 12 and 13
  spicata, 15, 16, 74, 91, 99, 111
  spicata var. *arizonica*, 3, 6, 11, 12, 92, 101, **103–107**, 225, 227, plate 13
  spicata var. *spicata*, 3, 6, 11, 12, 92, **103–107**, 225, 227, plate 14
  warnockii, 3, 10, 11, 12, 14, 15, 36, 92, 94, 96, 107, **108–112**, 221, 225, plate 14
hider of the north. See *Calypso bulbosa* var. *americana*
hooded ladies' tresses. See *Spiranthes romanzoffiana*

Ibidium
  strictum, 204
  romanzoffianum, 204

jewel orchids. See *Goodyera*
Juniperus deppeana, 5, 214

ladies' slippers. See *Cypripedium*
ladies' tresses. See *Spiranthes*
large coralroot. See *Corallorhiza maculata*
large yellow lady's slipper. See *Cypripedium parviflorum*
lesser rattlesnake plantain. See *Goodyera repens*
Lilium philadelphicum, 66
Limnorchis, 147, 148, 155, 171
  aggregata, 173
  arizonica, 164
  brevifolia, 154
  huronensis, 158
  media, 162
  purpurascens, 168, 171, 172
  sparsiflora, 173
  zothecina, 177
Liparis, 123
Listera, 17, **113**
  convallarioides, 3, 4, 11, 12, 15, 113, **114–117**, 119, 126, 225, plate 15
  cordata, 3, 5, 10, 11, 12, 13, 16, 25, 32, 51, 113, **118–121**, 162, 227, plate 15
  cordata var. *nephrophylla*, 118, 119
  nephrophylla, 118

Listeria, 113
long bracted orchid. See *Coeloglossum viride* var. *virescens*

Madrean adder's mouth. See *Malaxis corymbosa*
Madrean ladies' tresses. See *Spiranthes delitescens*
Malaxis, 8, 10, 17, 41, 57, 110, **123**, 127, 139, 156
  abieticola, 3, 5, 11, 12, 14, 15, 16, 117, 123, **124–127**, 134, 139, 167, 189, 225, 227, plate 16
  carnosa, 139
  corymbosa, 3, 4, 11, 12, 13, 15, 107, 111, 123, **128–131**, 134, 139, 167, 189, 225, plates 17 and 18
  ehrenbergii, 135
  macrostachya, 139
  montana, 136, 139
  porphyrea, 3, 5, 11, 12, 13, 15, 123, 126, 130, **132–135**, 139, 156, 167, 189, 225, 227, plates 18 and 19
  purpurea, 132
  soulei, 3, 5, 6, 11, 12, 14, 25, 46, 83, 101, 107, 111, 117, 123, 126, 130, 134, **136–139**, 156, 167, 171, 189, 215, 225, 227, plate 19
  tenuis, 127
  wendtii, 135
many flowered coralroot. See *Corallorhiza maculata*
Menzies's rattlesnake plantain. See *Goodyera oblongifolia*
Microstylis, 139
  corymbosa, 129
  montana, 136
  purpurea, 132, 135
  porphyrea, 132, 135
  tenuis, 124, 127
Milla biflora, 214
Mimulus
  cardinalis, 179
  guttatus, 179
moccasin flower. See *Cypripedium*
Montolivaea unalaschcensism, 143
Mycetophilidae, 120
mountain malaxis. See *Malaxis soulei*

Neottia
  michuacana, 211

striata, 43
net leaf. See *Goodyera repens*
northern coralroot. See *Corallorhiza trifida*

Oidaematophorus, 145
Orchiastrum romanzoffianum, 204
Orchidaceae, 1
Orchidinae, 27
Orchis
  huronensis, 158
  hyperborea, 153
  virescens, 28
Ophrys
  convallarioides, 114
  cordata, 118
  nephrophylla, 118

pale coralroot. See *Corallorhiza trifida*
Paphiopedilum, 59, 212
Peramium
  decipiens, 80
  giganteum, 70
  menziesii, 80
  ophilides, 85
  repens, 85
  repens var. *ophilides*, 85
Pinus contorta, 50
Piperia, 17, 27, **141–142**
  elegans, 141
  unalascensis, xii, 3, 10, 11, 12, 13, 16, 78, 141, **143–146**, 221, 227, plate 20
Platanthera, 4, 7, 8, 10, 17, 27, 117, 121, 144, **147–149**, 155, 156, 160, 161, 163, 165, 169, 170, 171
  aquilonis, 3, 10, 11, 12, 148, 149, **150–153**, 159, 160, 161, 162, 169, 170, 182, 219, 227, plate 21
  bracteata, 28
  brevifolia, 3, 10, 11, 12, 148, **154–157**, 166, 170, 182, 221, 227, plate 22
  dilatata, 16, 148, 160, 162, 163, 169, 171, 219, 221
  dilatata-hyperborea complex, 147
  dilatata var. *albiflora*, 148, 172
  dilatata var. *leucostachys*, 148
  flava var. *herbiola*, 30
  foetida, 143
  huronensis, 3, 10, 11, 12, 32, 67, 148, 149, 151, 152, **158–163**, 182, 227, plates 22 and 23

Platanthera (continued)
  hyperborea, 148, 150, 151, 153, 162, 171, 219
  hyperborea var. *huronensis*, 158
  hyperborea var. *purpurascens*, 168, 172
  limosa, 3, 8, 11, 12, 15, 41, 57, 126, 130, 134, 149, **164–167**, 182, 225, 227, plate 23
  obtusata, 222
  purpurascens, 3, 11, 12, 25, 30, 32, 41, 46, 51, 57, 67, 83, 88, 144, 148, 149, 151, 152, 166, **168–172**, 182, 189, 210, 225, 227, plate 24
  sparsiflora, 11, 12, 74, 148, 155, 156, 157, **173–176**, 178, 180, 182, 225, 227, plate 25
  sparsiflora var. *brevifolia*, 154
  sparsiflora var. *ensifolia*, 174, plate 25
  stricta, 169, 170, 172, 219
  unalaschcensis, 143
  viridis, 28
    × *correllii*, 172
    × *media*, 158, 162
  zothecina, 3, 11, 12, 15, 74, 148, 155, 176, **177–180**, 182, 222, 225, plate 26
Platyptilia, 145
Phithyrus, 23
Phragmipedium, 59
purple-spike coral root. See *Hexalectris warnockii*
purple malaxis. See *Malaxis porphyrea*
Pyrobombus, 23

rat-tailed malaxis. See *Malaxis soulei*
rattlesnake plantain. See *Goodyera*

satyr orchid. See *Coeloglossum viride* var. *virescens*
Satyrium
  repens, 85
  viride, 29
Schiedeella, 17, **185**, 193, 210
  arizonica, 3, 5, 8, 10, 11, 12, 15, 25, 32, 41, 46, 57, 84, 86, 117, 130, 134, 139, 156, 167, 171, **186–191**, 225, 227, plates 27 and 28
  fauci-sanguinea, 190, 191
  michuacanum, 210, 211
  parasitica, 190
  romeroana, 189
  violacea, 190

Schiedeellopsis, 185
Sciaridae, 120
Selenipedium, 59
Serapias
  gigantea, 70
  helleborine, 75
  latifolia, 75
shining cock's comb. See *Hexalectris nitida*
slender spire orchid. See *Piperia unalascensis*
sparsely flowered bog-orchid. See *Platanthera sparsiflora*
Spiranthes, 4, 16, 17, 185, **193**, 209, 210
  cernua, 201, 219
  decipiens, 80
  delitescens, 3, 11, 12, 15, 194, **195–199**, 225, plate 29
  diluvialis, 194, 196, 198, 221
  graminea, 196, 198
  magnicamporum, 3, 11, 12, 13, 16, 194, **200–203**, 205, 219, 227, plate 30
  mechoacana, 211
  nebulorum, 196
  ochroleuca, 201
  odorata, 201
  parasitica, 188, 190
  romanzoffiana, 3, 7, 8, 11, 12, 15, 88, 153, 194, 201, **204–208**, 225, 227, plate 31
  stricta, 204
  unalascensis, 143
  vernalis, 220
Spiranthinae, 185, 209
spring coralroot. See *Corallorhiza wisteriana*
spotted coralroot. See *Corallorhiza maculata*
squirrel ear. See *Goodyera repens*
Stenorhynchus michuacanus, 211
Stenorrhynchos, 17, 193, **209–210**
  michuacanum, 3, 6, 8, 11, 12, 13, 14, 15, 139, 210, **211–217**, 222, 225, plate 32
Stenorrhynchus, 209
Stenorynchus, 209
stream orchid. See *Epipactis gigantea*
striped coralroot. See *Corallorhiza striata*
Syrphidae, 72

tall green bog-orchid. See *Platanthera huronensis*
tall northern green orchid. See *Platanthera huronensis*
tessellated lesser rattlesnake plantain. See *Goodyera repens*
Texas purple spike. See *Hexalectris warnockii*
Thurber's bog orchid. See *Platanthera limosa*
Tipularia, 19
Triorchis
  romanzoffiana, 204
  stricta, 204
twayblade. See *Listera*

Vaccinium oreophilum, 66
Venus' slipper. See *Calypso bulbosa* var. *americana*
Veratrum californicum, 166, 170

whippoorwill-shoe. See *Cypripedium parviflorum* var. *pubescens*
white blotched rattlesnake plantain. See *Goodyera repens*
Wister's coralroot. See *Corallorhiza wisteriana*

yellow lady's slipper. See *parviflorum* var. *pubescens*

DATE DUE

DUE DATE SUBJECT TO CHANGE
IF A RECALL IS REQUESTED